ANALOG
ESSAYS ON
SCIENCE

OTHER RELATED WILEY SCIENCE EDITIONS

The Search for Extraterrestrial Intelligence, by Thomas R. McDonough

Seven Ideas that Shook the Universe, by Bryon D. Anderson and Nathan Spielberg

Space: The Next Twenty-Five Years, Revised Edition, by Thomas R. McDonough

The Body In Time, by Kenneth Jon Rose

Clouds in a Glass of Beer, by Craig Bohren

The Complete Book of Holograms, by Joseph Kasper and Steven Feller

The Scientific Companion, by Cesare Emiliani

Starsailing, by Louis Friedman

Mirror Matter, by Robert Forward and Joel Davis

Gravity's Lens, by Nathan Cohen

The Beauty of Light, by Ben Bova

Cognizers: Neural Networks and Machines that Think, by Colin Johnson and Chappell Brown

Inventing Reality: Physics as Language, by Bruce Gregory

Planets Beyond: Discovering the Outer Solar System, by Mark Littmann

The Starry Room, by Fred Schaaf

Ozone Crisis: The 15-Year Evolution of a Sudden Global Emergency, by Sharon Roan

Serendipity: Accidental Discoveries in Science, by Royston Roberts

The Starflight Handbook: A Pioneer's Guide to Interstellar Travel, by Eugene Mallove and Gregory Matloff

ANALOG ESSAYS ON SCIENCE

EDITED BY

Stanley Schmidt

WILEY SCIENCE EDITIONS
JOHN WILEY & SONS, INC.
New York • Chichester • Brisbane
Toronto • Singapore

This publication is designed to provide accurate and authoritative information in regard to the subject matter covered. It is sold with the understanding that the publisher is not engaged in rendering legal, accounting, or other professional service. If legal advice or other expert assistance is required, the services of a competent professional person should be sought. FROM A DECLARATION OF PRINCIPLES JOINTLY ADOPTED BY A COMMITTEE OF THE AMERICAN BAR ASSOCIATION AND A COMMITTEE OF PUBLISHERS.

Copyright © 1990 by Davis Publications, Inc.

Published by John Wiley & Sons, Inc.
All rights reserved. Published simultaneously in Canada.

Reproduction or translation of any part of this work beyond that permitted by section 107 or 108 of the 1976 United States Copyright Act without the permission of the copyright owner is unlawful. Requests for permission or further information should be addressed to the Permission Department, John Wiley & Sons, Inc.

Library of Congress Cataloging-in-Publication Data

Analog essays on science/ edited by Stanley Schmidt.
     p. cm. -- (Wiley science editions)
     Includes bibliographical references.
     ISBN 0-471-50839-X
    1. Science. 2. Technology. I. Schmidt, Stanley. II. Analog.
  III. Series.
Q158.5F763 1990   89-37010
500--dc20   CIP

Printed in the United States of America
90 91 10 9 8 7 6 5 4 3 2 1

ACKNOWLEDGMENTS

GRATEFUL acknowledgment is made to the following for permission to reprint their copyrighted material:

Base Eight Arithmetic, Meteors and Man by John Gribbin, copyright © 1981 by Davis Publications, Inc., reprinted by permission of the author; *Science & Creation* by Poul Anderson, copyright © 1983 by Davis Publications, Inc., reprinted by permission of Scott Meredith Literary Agency, Inc.; *The Long Stern Chase: A Speculative Exercise* by Rick Cook, copyright © 1986 by Davis Publications, Inc., reprinted by permission of the author; *Advanced Machining in Ancient Egypt?* by Christopher P. Dunn, copyright © 1984 by Davis Publications, Inc., reprinted by permission of the author; *A Little More Pollution, Please!* by George W. Harper, copyright © 1986 by Davis Publications, Inc., reprinted by permission of the author; *Demythologizing the Black Hole* by Richard Matzner, Tsvi Piran, and Tony Rothman, copyright © 1980 by Davis Publications, Inc., reprinted by permission of the authors; *Death Risk* by Milton A. Rothman, copyright © 1980 by The Condé Nast Publications, Inc., reprinted by permission of the author; *Hot Rocks & Water* by Richard Patrik Terra, copyright © 1985 by Davis Publications, Inc., reprinted by permission of the author; *Memetics and the Modular Mind* by H. Keith Henson, copyright © 1987 by Davis Publications, Inc., reprinted by permission of the author; *Man's Biological Future* by L. Sprague deCamp, copyright © 1980 by Davis Publications, Inc., reprinted by permission of the author; *Xenology: The New Science of Asking "Who's Out There?"* by David Brin, copyright © 1983 by Davis Publications, Inc., reprinted by permission of the author; *Alien Sex* by Robert A. Freitas, Jr., copyright © 1982 by Davis Publications, Inc., reprinted by permission of the author; *New Communications Technologies and the Developing World* by Arthur C. Clarke, copyright © 1982 by Davis Publications, Inc., reprinted by permission of Scott Meredith Literary Agency, Inc.; *Huntington's Handle* by Mark E. Peeples, copyright © 1987 by Davis Publications, Inc., reprinted by permission of the author; *Space Tourism: The Door into the Space Age* by Patrick Collins, copyright © 1988 by Davis Publications, Inc., reprinted by permission of the author; *Exploring the Asteroids* by Joel A. Davis, copyright © 1982 by Davis Publications, Inc., reprinted by permission of the author; *The Postdiluvian World* by Stephen L. Gillett, copyright

© 1985 by Davis Publications, Inc., reprinted by permission of the author; *Nanotechnology* by Chris Peterson and K. Eric Drexler, copyright © 1987 by Davis Publications, Inc., reprinted by permission of the authors; *24th Century Medicine* by Thomas Donaldson, copyright © 1988 by Davis Publications, Inc., reprinted by permission of the author; *To the Stars!* by Gordon R. Woodcock, copyright © 1983 by Davis Publications, Inc., reprinted by permission of the author.

CONTENTS

INTRODUCTION, *Stanley Schmidt* 1

        PART ONE. NEW VIEWS OF THE PAST 5

BASE EIGHT ARITHMETIC, METEORS AND MAN, *John Gribbin* 7

SCIENCE AND CREATION, *Poul Anderson* 15

THE LONG STERN CHASE: A SPECULATIVE EXERCISE, *Rick Cook* 24

ADVANCED MACHINING IN ANCIENT EGYPT?, *Christopher P. Dunn* 34

A LITTLE MORE POLLUTION, PLEASE!, *George W. Harper* 49

        PART TWO. THE UNIVERSE WE LIVE IN 59

DEMYTHOLOGIZING THE BLACK HOLE, *Richard Matzner, Tsvi Piran, and Tony Rothman* 61

DEATH RISK, *Milton A. Rothman* 84

HOT ROCKS AND WATER, *Richard Patrik Terra* 97

        PART THREE. WHAT IS THIS THING CALLED MAN? 109

MEMETICS AND THE MODULAR MIND, *H. Keith Henson* 111

MAN'S BIOLOGICAL FUTURE, *L. Sprague de Camp* 124

        PART FOUR. THE SEARCH FOR EXTRATERRESTRIAL INTELLIGENCE 137

XENOLOGY: THE NEW SCIENCE OF ASKING "WHO'S OUT THERE?," *David Brin, Ph.D.* 139

ALIEN SEX, *Dr. Robert A. Freitas, Jr.*   159

PART FIVE. COMING SOON...   167

NEW COMMUNICATIONS TECHNOLOGIES AND THE DEVELOPING WORLD, *Arthur C. Clarke*   169

HUNTINGTON'S HANDLE, *Mark E. Peeples, Ph.D.*   179

SPACE TOURISM—THE DOOR INTO THE SPACE AGE, *Patrick Collins*   193

EXPLORING THE ASTEROIDS, *Joel A. Davis*   205

THE POSTDILUVIAN WORLD, *Stephen L. Gillett, Ph.D.*   222

NANOTECHNOLOGY, *Chris Peterson and K. Eric Drexler*   236

PART SIX. BEYOND TOMORROW: THE FAR FUTURE   249

24TH CENTURY MEDICINE, *Thomas Donaldson*   251

TO THE STARS!, *Gordon R. Woodcock*   268

INTRODUCTION

BY STANLEY SCHMIDT

PEOPLE sometimes suppose that "science" and "science fiction" are about as different as two things can be. Consider, for example, the expression, "Oh, that's just science fiction!" commonly used to dismiss something regarded as impossible nonsense. I heard it often when I was growing up, applied to such wildly fanciful notions as flights to the Moon and organ transplants—both of which, as almost everyone knows, are now either past history or standard practice. (There have not been any flights to the Moon for a while, but there *is*, as I write this, at least one ship on its way to Mars, and others somewhere beyond the Solar System. . . .)

The reality is that there has long been a close, symbiotic relationship between science and science fiction. Science fiction, please note, is something quite distinct from fantasy. Both deal with happenings that may seem impossible to many of their readers, and fantasy makes no attempt to pretend that they are anything else. Science fiction at its best, on the other hand, strives to imagine things that may sound fantastic but also represent real possibilities. Sometimes these things are firmly rooted in existing science, as in the meticulously imagined worlds of Hal Clement or Poul Anderson. Sometimes they appear to contradict existing theory, as in the case of faster-than-light travel—but reality has forced us to revise theory before and will very likely do so again. A good science fiction writer can look at the seemingly impossible and imagine a way that it may turn out to be possible after all.

The type of mind that likes that sort of challenge is likely to be attracted to either research or science fiction writing—or sometimes both. Many science fiction writers, such as Robert A. Heinlein, Isaac Asimov, and Gregory Benford, have been scientists or engineers themselves. Many scientists and engineers went into those professions because their imaginations were fired by ideas they found in science fiction when they were young. It is no exaggeration to say that the exploration of space became a reality largely because certain young men in the 1930s and 1940s read stories that gave them the idea that it *might* be possible to go to the Moon, so they grew up figuring out how to do it.

The symbiotic relationship between science and science fiction was especially pronounced in one magazine, *Astounding Science Fiction*. *Astounding* was founded in 1930, but burst into full bloom in 1938 when its editorship was taken over by John W. Campbell, Jr. By professional training, Campbell was a physicist; by temperament and interest, he was a student of the entire universe, with high hopes for a creature called Man. He revolutionized science fiction by demanding, for the first time, equal emphasis on the *science* and the *fiction* in the stories he published. He made it a point to keep abreast of new developments in all fields of science and technology, and he goaded his writers into thinking about where those developments might lead. He encouraged writers and readers alike to question everything and, in particular, to beware of areas where science was threatening to fossilize and impose limits where maybe limits did not have to exist. By doing so, he helped keep scientists on their toes.

Not surprisingly, the devoted audience Campbell built for *Astounding* was as keenly interested in real science as in science fiction. Many of the readers were scientists or engineers, and they played "The Game" by pouncing fiercely on any author they caught slipping into scientific nonsense. Moreover, they devoured the fact articles as eagerly as the stories.

From the beginning, Campbell included at least one article of scientific fact in virtually every issue of *Astounding*. He considered the relationship between science fiction and science fact so close and so important that in 1960 (he remained editor for over 33 years, until his death in 1971) he changed the magazine's name to *Analog Science Fiction/Science Fact*. Under that title it still flourishes. The present editor is a physicist who grew up reading *Astounding*, began writing for *Analog* under Campbell's tutelage while in graduate school, and is determined to maintain that close, vital link between science and fiction in its pages. Now, as then, every issue contains a fact article, usually about some topic on the frontiers of research with far-reaching implications for the future.

This book is a sampling of those articles from *Analog* in the 1980s. Browsing through the table of contents, you will see that they cover quite a wide range of subject matter. All of them attempt to deal solidly and accurately with established factual material, but they seldom limit themselves to reviews of old knowledge. Many go beyond mere reporting of recent developments to informed speculation about what the future may hold—or how our present views of the past may need to be reinterpreted and refined. Many of the authors, such as John Gribbin, Poul Anderson, Arthur C. Clarke, and David Brin, are well known in both scientific and science-fictional circles. Some of them, such as K. Eric Drexler, are the people actually doing the pioneering work in the fields they write about. All, I think, will not only inform, but also entertain and stimulate your imagination.

I have arranged these essays in several groups that seem to me to form a logical reading sequence, though you needn't feel obliged to read them in that order. Those in Part One, "New Views of the Past," deal in one way or another with possible reinterpretations of our own history, from the very ancient to the fairly recent. Our knowledge of times before our own is, after all, based on rather sketchy evidence. Different ways of filling in the gaps could lead to quite different, but equally plausible, pictures of what really happened. Maybe we really do owe our present position at the top of the evolutionary heap to an astronomical calamity that ruined the dinosaurs' chances. Maybe the real explanation of the Egyptian pyramids is both closer to home and, in its way, more fantastic than extraterrestrial help.

Part Two, "The Universe We Live In," explores our physical environment on several levels, from the everyday (but surprising) business of risk assessment, to the changing picture of our Solar System that is emerging from the exploration of space, to one of the most fascinating (and popularly misunderstood) beasts in the astronomical menagerie. Part Three, "What Is This Thing Called Man?" considers some questions about our own nature, present and future. Part Four, "The Search for Extraterrestrial Intelligence," speculates about the other intelligences with whom we may be sharing the universe—and why we haven't met them yet.

Part Five, "Coming Soon," is just that: a look at some developments that appear likely in our own lives or those of our very near descendants. They range from new forms of communication and medicine, to the commercial development of the Solar System, to nanotechnology—an emerging field of real technology whose potential promises and threats leave most science fiction in the dust.

And beyond that? Well, nanotechnology is the bridge to the kind of medicine envisioned in Part Six, "Beyond Tomorrow." As the title suggests, these closing essays look a bit further down the road, to a time when death may be just another treatable condition and our kind may indeed have gone to the stars. There are still those who think we're kidding when we write about such possibilities—but, then, they thought we were kidding about the Moon, too.

If you're the type who understands that we weren't, and aren't, kidding, then I think you'll enjoy these articles. And I hope that you'll look for the new ones appearing every month in the latest pages of *Analog*.

Part One

NEW VIEWS
OF THE
PAST

John Gribbin

BASE EIGHT ARITHMETIC, METEORS AND MAN

DECEMBER 1981

WHY the dinosaurs became extinct has long been a favorite subject for speculation among scientists and laymen alike. The topic tickles people's imaginations, largely because of the kind of creatures the dinosaurs were, but does it really have any direct relevance for human beings? In a synthesis of ideas from several sciences, John Gribbin shows that it may indeed have such significance, of the largest possible kind. We may literally be here only because something happened to get the dinosaurs out of our ancestors' way—and the same thing could happen to us.

Dr. Gribbin holds a Ph.D. in astrophysics and makes his living by writing nonfiction books about science. The latest, *The Omega Point* (Bantam), discusses the ultimate fate of the universe and the nature of time. He has won several awards, both in Britain and in the United States, for his scientific writing. He also broadcasts on matters scientific for the BBC and occasionally ventures into science fiction.

The ideas discussed in this article are developed further in his book *Children of the Ice* (Blackwell). ∎

WHY does our arithmetical system use the base ten? Obviously, it is because we have two hands each, and each of those hands has five digits on it. There is nothing sacred about base ten arithmetic, and if, with a little imagination, we envisage an intelligent life form with four hands, each having three fingers, then logically we might expect that life form to count in twelves. On the face of it, assuming that intelligent beings start to count by using convenient parts of their manipulating limbs as markers, there are endless possibilities for other life forms to base their arithmetic on.

But how many of these are actually practical possibilities? To start with, would it really make sense to have four arms, each with three fingers? In evolutionary terms, probably not—at least as far as intelligent life is concerned. Bilateral symmetry, with limbs paired on either side of the body, is clearly a successful evolutionary invention. Legs on each side help you to stand up, and an odd number of true legs cannot be conveniently fitted into this pattern; although some monkeys have a grasping tail, and the kangaroo uses its tail as a prop. Once legs to stand on and move about with have been "invented," it is a logical evolutionary step to adapt some of those limbs for manipulating things. In our case, this means that the other pair has been adapted for walking on, which just leaves the two hands free. We should be cautious about reading too much into this, since, after all, our cousin-apes, the chimpanzees, make considerable use of their "feet" for grasping, and would probably regard base 20 as "normal" if they were bright enough to invent arithmetic. But *really* accurate manipulation involves good eye/brain/hand coordination. That is now thought to be a major reason why the human brain developed, and such coordination works best with the limbs nearest to the eyes, at the front —or top—end of the body.

There is also a principle called "maximum parsimony"—a variation on Occam's Razor—which says that it is advantageous in evolutionary terms to make the minimum adaptation needed for success. An extra pair of arms might be useful occasionally, but the occasional advantage has to be offset against such mundane aspects of everyday life as the need to eat more food to provide for those arms and the doubled risk of breaking a manipulating limb and being disabled. Whatever the reasons, all the successful species on Earth which go in for manipulating limbs— even species with six or eight limbs—always set aside just one pair for the purpose. The crab has two large pincers at the front, the kangaroo has two arms, and even a mouse uses its front pair of limbs for holding food. This makes sense. With bilateral symmetry and eyes at one end of the body, obviously one pair of limbs will be most useful for grasping and moving things.

Leaving aside wild speculation about life in the clouds of Jupiter or on the surface of a neutron star, we might begin, on this basis, to decide just how far our own shape is determined by the conditions under which our ancestors evolved. Are we, in fact, typical of the kind of intelligent life to be found on Earth-like planets? What are the chances that, if ever we do make contact with intelligent beings that have evolved under similar conditions, they too will be upright, bipedal animals with two arms, each ending in a five-fingered hand, and a head mounted on top of the body carrying a pair of eyes, a nose, and a mouth?

To start with, intelligent life—the kind that builds civilizations and spaceships, that is—can't be otherwise too successful in being adapted to its own natural environment and must have been, during its evolution, under considerable pressure from predators. The elephant is intelligent, by some standards, but so powerful that it is insulated from the dangers of attack by enemies and has never had to use its intelligence to fight off enemies. The whale and dolphin, potentially as intelligent as man, also have few enemies and are superbly adapted to their watery environment. The price they have been obliged to pay is streamlining and a total absence of limbs and hands that can be used to manipulate objects. A whale may sing, but he can neither construct nor play the cetacean equivalent of a saxophone or piano.

The point is not unimportant. Intelligent, tool-using life will emerge on a planet like ours on the land, not in the sea. It won't be very big or fierce, since big, fierce animals have no need to invent tools or weapons, or to sharpen their cunning by hiding from predators. And it will have a pair of limbs at one end, conveniently near the eyes and ending in digits (fingers) that can be used to grasp and manipulate small objects.

The picture already begins to look very much like a primate: a ratlike or squirrellike creature good at hiding and scurrying out of danger, with sharp eyes and good hearing to detect danger coming, and not so big that hiding or running away becomes difficult. What about more legs for running away with? The centaur, half horse and half man, looks at first like quite a good bet. But there is a snag. The bigger the body, the harder it is to hide and the more food it needs to survive. A centaur is heavily committed to running, rather than hiding, as a defense, and in evolutionary terms that means that the pressures of natural selection will operate to produce more horse-like centaurs, with the human-like limbs withering away into ever more useless appendages. No, apart from the kangaroo's tail, it is hard to see how we could improve on the basic design of two legs for running with, two arms with hands for carrying things, and a head mounted on top carrying two eyes to provide a stereoscopic, three-dimensional view of the world. Three-dimensional vision is essential for judging distances, whether it's the distance to a charging lion or to a morsel of food waiting to be picked up; a good high mounting for the eyes is essential for a prey animal, which needs an early warning of impending danger. The necessity of air to breathe and food to eat completes the outline design, requiring a mouth and a nose of some description, although maybe a few variations on those themes are possible.

At a quick glance, even trying to avoid any cultural bias from our everyday experience, it looks very much as if the bipedal design is the right one for intelligent life on Earth. The only real room for variation is in the number of fingers on each hand. Five is certainly a useful number, as we all know. But it does seem to be on the high side of usefulness. It is difficult to see how an extra finger on each hand would be very beneficial, while, by contrast, many people who have suffered accidents leading to amputations manage very well with only two or three fingers on a hand. The key, in those cases, is that they still have a thumb with which to oppose the remaining fingers, making it possible to grip and manipulate objects dexterously.

So far, all this is speculation. As yet, we have no information about life on other planets with which to test the idea that intelligent Earth-type life is bipedal and, broadly speaking, manlike. What we need is one test case. If we landed one robot probe on one other Earthlike planet and found the dominant life form to be an intelligent biped with four or five fingers on each hand, the argument that this is the inevitable product of evolutionary selection on such a planet would be overwhelming. The chance of such similarities arising by coincidence is so small as to be virtually negligible.

Unfortunately, the chance of landing a robot probe on another Earthlike planet in the immediate future is equally small. But wait—this isn't the end of the story.

We do have information about one planet under conditions that were undeniably Earthlike but on which there was no human life. I refer, of course, to the Earth itself during the era of the dinosaurs. If the arguments I have sketched out above hold water at all, then the pressures of natural selection, operating during the era of the dinosaurs just as they have in the subsequent 65 million years of Earth history, should have been acting to produce an intelligent, upright biped.

Now, during the 150 million years or so that dinosaurs dominated the Earth, the evolutionary pressures were, in many ways, less than they have been since. In particular, the climate was more stable than it has been in the past few million years, and because of the geographical arrangement of the continents (which changes over millions of years due to continental drift), there were no great Ice Ages to weed out species and put a premium on intelligence and adaptability. The recent cycle of Ice Ages, according to most evolutionary theorists, played a key part in forcing man to adapt to changing conditions, putting a premium on intelligence and flexibility, and making us what we are today. That is why man has evolved so rapidly.

But even with less pressure on the dinosaurs from the environment, surely intelligence would still be an advantage? And surely over 150 million years even relatively gradual evolutionary changes would have had a chance to get to work?

Indeed they would. Although most people think of dinosaurs as great lumbering brutes with tiny brains, in fact the term applies to a variety of creatures as wide as the variety covered by the term mammals today. There were big, stupid dinosaurs, but there were also small, agile dinosaurs. There were meat-eaters—the dinosaur equivalent of lions and tigers—and there were grass eaters—the dinosaur equivalent of deer and sheep. Think of any variation on the mammal theme today, and the chances are that there was a dinosaur equivalent. And the dinosaurs didn't even die out without a trace, whatever the impression most popular accounts provide. Dinosaur descendants are alive and well on Earth today: not just in the form of obviously reptilian creatures like crocodiles and alligators, but also in the form of birds, products of a highly successful dinosaur line that took to the air (as well as developing warm blood, a trick emulated by other dinosaur lines). Out of all that variety, were there no dinosaur candidates for the bipedal, upright niche that, according to my argument, marks a vital place on the road to intelligence? If the fossil record showed no sign of a dinosaur even remotely human in appearance, we would have to admit that the idea falls down; but if there were dinosaurs that could be described, in the broadest terms, as on the path to humanlike appearance, then it would at least make the argument look a little more plausible.

In fact, there were several dinosaur types which followed, broadly speaking, the kangaroo's approach to bipedalism rather than the human approach, keeping a large tail which could be used as a stabilizer, weapon, or seat. That no problem; a biped with a tail is still a biped. Tyrannosaurus and Iguanodon carried this design to extremes, reaching five meters in height. One a flesh-eater, the other herbivorous, neither of them could be said to be intelligent. Scleromochlus, a bipedal reptile about one meter long, which lived about 200 million years ago, is superficially a more likely candidate for the pre-intelligent niche, but it had a small brain and never seems to have made the grade. But there is a star candidate who fulfills, as far as we can tell from the fossil remains, all of our requirements. If you landed on a distant planet and were greeted by a creature like Saurornithoides, you would have to admit that the argument that Earthlike planets produce manlike intelligent species holds water.

A Saurornithoides was a smallish dinosaur, weighing about 50 kilos, which lived at the end of the age of the dinosaurs, some 65 million years ago. It had the largest brain, in proportion to body mass, of any dinosaurs, with a brain-to-body weight ratio not far different from that of the modern baboon. And it was clearly an active bipedal creature, with a long tail behind and four-fingered hands at the end of each arm, the fingers perhaps being arranged as two true "fingers" with an opposable "thumb" on either side.

This is a pretty impressive set of credentials. Starting from this basis 65 million years ago, if Saurornithoides had followed the same path, in response to similar evolutionary pressures, that the equivalent pre-humans were to follow 60 million years later, then it might well have been possible for a Saurornithoides civilization to arise, with eight-fingered, kangaroo-like bipeds developing spaceflight by about 60 million years ago. If so, and if the species had survived whatever unimaginable processes lay in the 60 million years they could have had beyond the present stage of human civilization, the solar system today might well be the playground of a bipedal society, but one to which base eight arithmetic seemed the obvious choice. Carl Sagan speculated briefly along these lines in his entertaining book *The Dragons of Eden*. But why did the dinosaurs fail to make the breakthrough to intelligence? What stopped Saurornithoides from exterminating the mammals and going on to develop its own civilization?

The best answer seems to be that a large meteorite struck the Earth just when these particular dinosaurs were making the first steps on the road to intelligence, and as a result all the large animals living on the Earth's surface died. This is the explanation for the catastrophe which brought an end to the age of the dinosaurs which is currently in vogue, and it rests upon some very good evidence.

The fact that there was a catastrophe which wiped out all large animals is clear from the geological record. Almost overnight in geological terms (which means in the space of no more than 100,000 years), half of all the land species on Earth, including all animals bigger than about 40 kilos in body mass, became extinct. Following the disaster, the world was a different place. The surviving small animals, in particular, were now free to move into the ecological niches previously occupied by the large dinosaurs. Most of the small animals moving into those niches were mammals—the small mammals were already well established on Earth during the age of the dinosaurs—and over 65 million years they have evolved into elephants, tigers, gazelles, and so on, replacing the dinosaur equivalents now just a fossil memory. If the disaster, whatever it was, had wiped out all animals bigger than 60 kilos, then Saurornithoides would have been well placed to achieve world dominance. As it was, it just missed the boat, and in the fullness of time the little ratlike mammals which had probably been among his prey produced a new intelligent species—but one very much, in terms of superficial appearance, in the Saurornithoides mold. Of course, we have no tails and we have five fingers on each hand. But we look as much like Saurornithoides as we do like the tree shrews from which we are descended. On Earthlike planets, it seems that the way to fit the niche for intelligent life is indeed to be bipedal with two arms, two hands, and a head mounted on top in the lookout position.

But it also helps to avoid large meteor impacts. The chances of winning this particular cosmic lottery are not very good, at least in our solar system.

The battered faces of the Moon, Mercury, and some of the moons of Jupiter and Saturn bear mute witness to the frequency of meteoric impacts during the history of the solar system. Even after the effects of erosion by wind and water, the surface of the Earth shows that such impacts are still hardly rare on any geological timescale. Barringer Crater in Arizona is the classic example. More than 3900 feet across and 600 feet deep, it was produced by a meteoritic impact which can be dated, using standard geological techniques, at only 25,000 years ago. Vastly greater features, such as the West Clearwater Lake in Quebec (13 miles across) and the Vredevoort Ring in South Africa (35 miles across) show the characteristic circular shape of a meteorite impact and are almost certainly craters produced hundreds of millions of years ago. Clearly, such impacts must have dramatic environmental effects, and it is now fifteen years since Joe Enever presented the now-classic calculation of the worst kind of meteoritic disaster, published in *Analog* ("Giant Meteor Impact," March 1966).

Enever started with simple calculations of the energy involved in producing the Vredevoort Ring, using one of the simplest equations in physics—a body of mass m moving at velocity v has a kinetic energy of $\frac{1}{2}mv^2$, and if that body is brought to a halt by colliding with the Earth, all that energy is liberated as heat.

A fairly ordinary meteorite might be moving at 50 km per second when it hits the Earth, and there are bits of such cosmic rubble around in the solar system with masses of thousands of tonnes. The kinetic energy released by an impact with such a body would yield the equivalent of more than 100,000 megatonnes of TNT—bigger than any nuclear device yet tested by man.

Even this, however, is not enough to explain the Vredevoort Ring, which required an impact yielding 10 million megatonnes' equivalent, coming from a collision with an object as big as the asteroid Hermes—32 thousand million tonnes of rock.

If such an object had struck our planet 65 million years ago, it could well explain the demise of the large dinosaurs. Dust blasted high into the stratosphere by the explosive impact would have spread around the Earth like a shroud, blocking out sunlight, killing the plants beneath to deprive animals of food, and perhaps starting an Ice Age, or at least a mini-Ice Age, to finish off the starving survivors.

The snag with the hypothesis is that there is no crater comparable to the Vredevoort Ring but only 65 million years old to be found on Earth. But, as Enever pointed out, most of the Earth's surface is covered by water. Suppose the giant meteorite fell in the sea?

It might seem that an oceanic strike would be less spectacular than one on land, since it would be "damped" by the water. But the opposite is the case! Quite apart from incidentals such as the tidal waves produced, the almost unimaginable amount of energy released by the impact would not only vaporize the water of the sea at the point of impact, but would punch a hole scores of kilometers wide right through the thin crust of the ocean floor, exposing the hot magma beneath. Seawater pouring into the pit would eventually cool the molten rock and return conditions to normal—but not before 16,000 cubic km of water, on Enever's calculations, had been evaporated in the process.

In this version of the scenario, the Earth would be shrouded by shiny white clouds, reflecting away the Sun's heat, and the water vapor would be precipitated as snow. Once again, plants would die in profusion and animals would starve, with the biggest animals, that need most food, suffering the worst.

Everything fits. But the idea remained a speculation until 1979, when a team from the University of California at Berkeley came up with evidence that geological strata 65 million years old are enriched by traces of heavy elements, in particular iridium. The original discoveries were made in strata from Italy; since then, fresh evidence has come in from as far afield as Denmark, New Zealand, and the central North Pacific. All the evidence suggests that some global event 65 million years ago—just at the time of the death of the dinosaurs—spread a layer of dust enriched with exotic heavy metals around the world.

The best candidate for such an event is a giant meteor impact. The Earth's crust is deficient in heavy metals because any present when the Earth formed have settled into the dense, molten core. But asteroids, the cosmic rubble left over from the formation of the solar system, presumably contain a higher proportion of elements such as iridium, since they have no cores into which heavy elements can settle. The traces found in the key strata are still only traces by any normal standard, but amount to enrichment of the natural level of iridium by between 10 and 1000 times. Clearly, something happened 65 million years ago, and it would be a remarkable coincidence indeed if that something were not related to the disasters that brought an end to the era of the dinosaurs. The paleontologists, traditionally a cautious crew, have so far only acknowledged that a giant meteor impact may have contributed to the demise of the dinosaurs, perhaps being the "last straw" that came after several million years of deteriorating climatic conditions. Whatever, there seems no doubt at all that an event like the one which produced the Vredevoort Ring, if it happened tomorrow, would certainly spell the end of our civilization, if not of the entire human species (among others).

If it did, though, I'd be willing to make a small hypothetical wager that in 50 or 100 million years' time there would be a species of intelligent bipedal animals doing very nicely on Planet Earth. They might not be mammals or reptiles; they might or might not have tails. Maybe they would count in base eight or base ten, but I'd be surprised if they counted in either base six or base twelve. They'd be about two meters tall, with eyes upon heads at the top of their bodies. And they'd be speculating about the disaster that brought an end to the age of the mammals, wondering whether the upright bipeds had ever achieved true intelligence, and no doubt joking about the likelihood that those strange creatures with five-fingered hands may have used a bizarre decimal counting system. *Plus ça change; plus c'est la même chose.*

Poul Anderson

SCIENCE AND

CREATION

SEPTEMBER 1983

THE concept of evolution was highly controversial when it was first formulated, but later became generally accepted. In recent years, it has again become a matter of heated debate, though not because of new scientific evidence that threatened to undermine it. The recent controversies have deeper philosophical roots than that, as well as important implications for our civilization. As usual when a subject arouses intense emotions, the heat of those feelings tends to interfere with rational discussion on both sides of the issue. In this essay, Poul Anderson attempts to bring rationality back to the debate.

Poul was trained as a physicist, but began selling stories to *Astounding* while still in college. He soon became so successful at that, that he made his career as a full-time writer rather than as a physicist and now enjoys a reputation as both one of the most prolific and one of the best writers in several fields. Particularly noted for his ability to create unique, highly realistic, and multidimensional worlds populated by multidimensional people, Poul has published about a hundred books and many, many short pieces. He has won seven Hugo awards and three Nebulas and has served as president of Science Fiction Writers of America. Writing is very much a family affair with the Andersons: Poul's wife Karen is also a writer, and their son-in-law, Greg Bear, is the current president of SFWA and generally considered one of the best of the younger writers. ■

ONE of the less endearing—and more dangerous—features of the 20th century has been a worldwide tendency to substitute rhetoric for discourse. By now, reasoned debate is a rarity. There is seldom even any effort to understand an opposing point of view. Instead, a person attributes opinions or attitudes to the other fellow and proceeds to heap billingsgate upon him because of them, although they may not actually be what he means at all. Living near Berkeley, California, I have over the years watched this sort of thing develop in the academic community and its hangers-on, until I am inclined to agree with a fictional character of mine who remarked, "Sure, I'm anti-intellectual. I prefer people who think."

Well, that may be just a little exaggerated, a hint of the same behavior I was condemning. It got your attention, though, didn't it? Let me try to make the rest of this essay an exercise in rationality.

I propose to discuss the "scientific creationism" which is so much in the current news. My conclusion will scarcely surprise you: that "scientific creationism" is a contradiction in terms. If that were all, there would be no point in stating it yet again. Why preach to the choir? However, it does seem to me that spokesmen for the scientific establishment have generally made their points poorly, because often they themselves don't quite realize what the concept of evolution signifies. Thus the argument we'll advance against creationism here will take a turn that may prove surprising, therefore enlightening to some readers. Indeed, it will be only the first step in a brief exploration of the philosophy of science.

We begin by foreswearing invective. The creationists are *not* a bunch of yahoos. They are generally well-educated and well-mannered individuals, a number of them with excellent scientific credentials. (While I don't know just what James Irwin's views on evolution are, we all know he believes the Biblical story of Noah is substantially true, and led an expedition in search of the remains of the Ark—after having been on the moon!) Nor do most of them want to suppress any other doctrine. Socially and politically, they have several quite valid, important points to make. Secular humanism has in fact become the teaching of the public schools, to the exclusion of crucial parts of our heritage. The effects on culture are already sad, the implications for the future of liberty and even for national survival ominous. Would it really infringe on anybody's constitutional rights if children were to learn something about the roots of their civilization?

But this does not mean they should learn things, at taxpayer expense, which simply are not true. By now, the scientific attitude and the body of discoveries to which it has led are themselves basic to society, and not merely Western society. "Scientific" creationism is not content to maintain that the universe is the work of God. It claims that this Earth is, at most, a few thousand years old, and that the species of living beings we know today came into being in their present forms. Of course, the First Amendment guarantees any American the right to believe and argue for that and teach it privately. But the notion has no more claim on "equal time" in public education than do, say, astrology, psionics, or Marxism.

It is scarcely necessary here to repeat what has often been pointed out: that if the creationist assertion is true, then our astronomy, physics, chemistry, biology, and archaeology must be false. For example, evidence for geological ages includes matters as diverse as the well-established laws of radioactive decay and a cosmic red shift observed by familiar techniques of spectroscopy. Much has been made of certain unexplained anomalies in certain mineral formations—far too much. Science is always coming upon such phenomena, and needing time and effort to learn what brings them about. We don't yet understand ball lightning very well, either; but nobody says that, on this account, we should throw out our meteorology. Instead, what understanding we do have provides a context within which to seek explanations of countless details.

Thus the claim that our planet is less than a million years old, and has undergone no significant changes during its existence, is incompatible with science. At best, a person might declare that God created the universe recently, full of misleading clues to something quite different. Emotionally, I am inclined to think this is an insult to the Creator. In the famous words of Einstein, "The Lord is subtle, but He is not malicious." Logically, we need only note that the declaration is, by its nature, untestable, incapable of being disproven; therefore it is devoid of empirical meaning.

We might, though, find it worthwhile at this point to refute one statement frequently made by creationists, that the development of matter and life from primitive to complex forms would violate the second law of thermodynamics. Even some who have accepted evolution as a fact, such as the late Lecomte de Noüy, have maintained that it would have been statistically impossible without supernatural guidance. They should have known better.

Part of the problem arises because the second law is deceptively simple-looking but has profound and far-reaching implications. It can be expressed in confusingly many ways, and has been. In one of my college textbooks (*Physical Chemistry*, by Frank MacDougall, Macmillan, 1944) the phrasing of the law goes: "It is impossible to devise any mechanism or machine by means of which a quantity of heat can be converted into the equivalent amount of work without producing other changes in the state of some body or bodies concerned in the process." Another book (*Introduction to Theoretical Physics*, by Leigh Page, Van Nostrand, 1928) puts it as: "No self-acting engine can transfer heat from a body of lower temperature to one of higher temperature." Here a "self-acting engine" means, essentially, one which is isolated from outside influences and which takes its working-substance through one or more complete cycles.

There are numerous other, equally valid versions of the same truth, but these two should be enough to show that we are dealing with something which is quite basic and not at all self-evident.

Not so incidentally, the first law of thermodynamics amounts to the law of the conservation of energy—that is, that the total amount of energy in an isolated system can be neither increased nor decreased. While the only completely isolated system is the universe as a whole, in practice we can make arrangements which come very close to that condition.

Were it not for the first law, we could build perpetual motion machines of the first kind, as they are called, whose driving power comes from nowhere. Were it not for the second law, we could build perpetual motion machines of the second kind, which would draw their energy from anywhere. For instance, a motor at room temperature could be driven by the air molecules that happen to collide with its piston. The impossibility of this is not immediately obvious, which is why it was not established until the 19th century and is still overlooked in many science fiction stories.

Numerous people who understand the second law as a principle governing heat engines can get bewildered about the wider applications. These involve entropy, about which science fiction has also perpetrated a great deal of nonsense. Actually entropy is a measurable quantity, though you need calculus to describe it mathematically. In any thermodynamic process where an amount of heat Q is exchanged at a temperature T (which may vary throughout the volume and during the time in which things happen), the increase of entropy is equal to the integral of dQ/T.

Now "increase" can be negative, that is, represent a decrease. When something occurs thermodynamically in a system, entropy can and often does decrease somewhere. However, it increases elsewhere, and the second law states that the *total* gain in entropy is always positive. That is, whenever a change involving an energy transfer takes place in a system, entropy always is greater at the end of the process than it was in the beginning.

A "system" can be anything: an atom, a molecule, a machine, a living organism, a galaxy, the cosmos as a whole, anything. But we must consider the entire system, not just a selected part of it.

An increase in entropy corresponds to, or measures, an increase in disorder, or a decrease in the orderliness of the system. Therefore, whenever something changes, we find there is less order afterward than there was before.

Here is a very rough example or analogy. Think of a house whose lady has brought it to absolutely perfect arrangement and cleanliness—not a single item of furniture out of place, not a speck of dirt or dust anywhere. Then her children come home from school and her husband home from the office, and the family starts using the place. Things *happen* in it. The immaculate condition doesn't last long, does it?

True, next morning the lady can restore her dwelling to its former orderliness. However, to do so she must expend energy, both her own and the energy of whatever appliances she uses, such as a vacuum cleaner. That energy comes from the conversion of food in her body—or fuel in an electric generator somewhere—into disorganized gases and masses. The house may become neat again, but the environment as a whole is more chaotic than it was.

This is not an argument against good housekeeping! It is simply a reminder that everything has its price.

The growth of life itself, and the maintenance of its exquisite complexity, is a wonderful example of order brought out of seeming chaos. Far too readily can we overlook the price paid, the net disorder. The solar energy which drives life was once tidily located in the sun. Its diffusion through the universe involves an enormous increase of disorder, of entropy. The profit that life makes along the way is minuscule by comparison.

In fact, even considered by themselves, biological processes require entropy increase. (The mean free energy does decrease, but that is an entirely different quantity.) We too are heat engines, subject to the same laws as all others.

The whole starry cosmos exemplifies the principle. There was no primordial chaos before the Big Bang—not really. Instead, everything was neatly concentrated in one location. Then it scattered, and is still scattering, a disorderliness far exceeding the structural order of galaxies, stars, planets, and life forms which have appeared in the course of the process.

Unfortunately, too few spokesmen for science grasp this themselves. Hence they are brought to incoherence by the specious claim that evolution is thermodynamically impossible. True, the refutation of that claim, even on the rudimentary level of this essay, might well drive away the popular audience whom the scientist wants to enlighten. Yet if the universe were any easier to describe than it is, would it not be that much the less miraculous?

In a still more subtle way, it seems to me that advocates of evolution put themselves at an unnecessary disadvantage by underrating the concept for which they are arguing. Here we begin to touch on the nature of science itself.

Creationists generally talk of the "theory of evolution." Many who disagree with creationism reply that evolution is no such thing, but a fact. Thereby, they fall into the same dogmatism as certain of their opponents, and become subject to the same refutation. After all, what is a "fact"? Nobody alive has ever met an Australopithecus or watched prokaryotic cells develop in the pre-Cambrian seas. It is a rather feeble retort that nobody has met Adam and Eve, either, or watched the world coming into being by fiat.

In the last analysis, those of us who accept the idea of evolution do so because it is an inference, based on many different accumulated observations, which enables us to account for those data, fit them into a scheme that makes sense. The creationist can quite legitimately reply that this is what his beliefs do for him.

However, at this point in the history of science, it is a mistake to agree that evolution is a mere "theory." That concedes more to the creationist than he deserves.

What is a theory, anyway? To answer that question, we must take a look at the scientific method itself.

Now, a number of distinguished scientists have denied that there is any such thing, and I rather agree with them. That, though, would take us too far afield now. Let us just glance at the traditional paradigm, oversimplified though it is. The purpose will only be to make clear what we mean by certain words.

In this paradigm, scientists begin by making observations of nature, as exact as possible. Then somebody formulates a scheme which summarizes those observations, preferably in mathematical terms. That is because mathematics is the language *par excellence* of precision. Somebody else takes such a *description* and tries to explain it by a *hypothesis*. That is, this person proposes the existence of a mechanism or a relationship which would logically produce the observations themselves. A good hypothesis also yields *predictions*; it tells us what further observations we should try to make. If we make them, and the results fit the scheme well enough, then in due course the hypothesis gains the status of a *theory*. That is, we accept it as depicting, more or less correctly, some aspect of reality.

Later discoveries may prove irreconcilable with the theory. In that case, we have to discard it—or, at least, drastically modify it—and look for another.

The standard example comes from planetary astronomy. For untold millennia, observers had been gathering data about the motions of the heavenly bodies across the sky. This effort culminated, for the time being, in the magnificent work of Tycho Brahe in the sixteenth century. Meanwhile, of course, there had been many attempts to account for the data. The idea that everything revolves around Earth grew increasingly unlikely as information accumulated; the picture had to be made too complicated, with epicycles. As early as the thirteenth century, Alfonso X, king of Leon and Castile, remarked that if he had been present at the Creation, he could have given the Creator some good advice!

Eventually Nicholas Copernicus offered a much more satisfactory description, in which the sun was at the center. Galileo Galilei and others refined this system and added to it. Finally Johannes Kepler put it into elegant mathematical form in his three laws of planetary motion.

Isaac Newton then accounted for those laws by his hypothesis of universal gravitation (even though he himself denied making hypotheses), together with his own three laws of the motion of all bodies, not just planets. Soon observation confirmed this so well that it became a basic theory in physics. By means of it, later generations discovered new planets and explained the behavior of distant stars.

There remained a few loose ends, such as a slow change in the orbit of Mercury. Early in the 20th century, Albert Einstein proposed a whole new theory, general relativity, which included Newtonian mechanics as a special case and which accounted for those anomalous phenomena.

Thus far, the usual description of science in action. As said, it is much oversimplified, and in many instances is scarcely true at all. Still, if nothing else, it does help us give clear meaning to our words.

The important point here, though, is that even taken at face value, it is incomplete. It omits a further stage of thought which is of primary importance.

Before going on to that, let us very sketchily review the history of the evolutionary concept. That way we can compare it to the development of astronomy. If nothing else, we will be reminding ourselves that the idea of evolution was not invented by a few subversives in the 19th century, but has a long and honorable past of its own.

By 1800 the concept was already in the air. There had been some speculation along those lines as far back as Classical times, if not before. During the Renaissance and after, men gradually realized that they were coming upon the petrified bones of beasts which no longer existed. Early in the 19th century, the great French naturalist Georges Cuvier advanced the hypothesis that more than one creation had occurred in the past: that life had appeared several times, to be wiped out by worldwide catastrophes, and that the account in the Bible refers only to the latest of these eras. Regardless of this deferral to religion, Cuvier was considered blasphemous by many. Once some of his students decided to throw a healthy scare into him. One of them costumed himself like the traditional Satan, entered the professor's home at night, woke him, and roared, "I am the Devil, and for your impiety I have come to eat you!" Cuvier looked him up and down and replied scornfully, "Hmf! Horns and hoofs. You can't. You're graminivorous."

His catastrophism was denied by a contemporary compatriot, Jean Baptiste Lamarck. A war hero at age sixteen, Lamarck later boldly maintained that living species had developed from less specialized ancestors. However, he thought that the causes lay in environment and the actions of individual organisms. This was so unconvincing that few accepted it until the 20th century, when for a time a version of it became official dogma in the Soviet Union.

In 1830 the Englishman Charles Lyell published the first volume of his epoch-making *Principles of Geology*, in which he showed that the forces that had shaped Earth in the past were the same as those at work today. By then it was becoming clear that man had coexisted with many animals long vanished, and in 1836 the Dane Christian Thomsen laid the foundations of modern archaeology by his scheme of successive Stone, Bronze, and Iron Ages, with the Stone Age reaching extremely far back in time.

Public as well as scientific interest in prehistory grew apace. More and more fossils were collected, in the Old and New Worlds alike, and reconstructions were made. When the Crystal Palace exposition opened in London in 1856, it included several life-sized statues of dinosaurs. Since they were not labeled, many visitors were puzzled by them. One man guessed that they were intended as an object lesson in temperance, to show what drunkards might expect to see.

In the same year, remains of Neanderthal man first came to light, in Germany. Initially, most biologists denied that this could be an extinct form of human, and various fanciful stories were devised to account for it. Yet evidence continued to accumulate, while the growth of geological knowledge made it less and less easy to believe that such creatures as the dinosaurs had perished in the Biblical flood—that they were, in the phrase of that day, antediluvian.

In 1859 Charles Darwin published *The Origin of Species*. This stunning demonstration of evolution as an understandable, natural set of processes—a hypothesis which had occurred independently, in less detail, to Alfred Russel Wallace—was followed four years later by another intellectual bombshell, Lyell's book *The Antiquity of Man Proved by Geology*. At the same time, field workers such as the Frenchmen Boucher de Perthes and Èdouard Lartet were turning up ever more traces of archaic humanity. When Darwin issued *The Descent of Man* in 1871, he did not "prove we are descended from apes." What he did was describe *how* humans and simians could have stemmed from a common ancestor; the idea that this *had* happened was, by then, current.

Of course, it had met with much opposition, both popular and scholarly. Southerners during the American Civil War were fond of saying that maybe Yankees came from monkeys, all right, but Mar'se Robert E. Lee couldn't be related to anything with a tail. Most clergymen combatted every suggestion that the Book of Genesis was not a straightforward piece of reporting.

In fairness, we must add that not all did; indeed, some made important contributions to knowledge in this field, especially in France. For that matter, Thomas Henry Huxley's debating opponent, Bishop Samuel Wilberforce, was by no means a bigoted ignoramus, but a cultivated and philanthropic gentleman.

Nonetheless, the data were accumulating remorselessly. In a paper read in 1865, the Austrian monk Gregor Mendel established the basis of genetics. His work went almost unnoticed for a generation, but came back to light after the Dutchman Hugo de Vries had identified the phenomenon of mutation, about 1895. Here was the decisive last factor that Darwin had not known of, the material on which his principles of natural selection and sexual selection operated. Meanwhile, in 1891 another Dutchman, Eugène Dubois, had found in Java the relics of a being that was unequivocally related to man and yet far too primitive, too apelike, to be *Homo sapiens*.

Meanwhile, too, knowledge was rapidly growing of the world as it had been long before anything like us existed. A clear-cut example is the evolutionary lineage of the horse, established through fossil finds by the American O.C. Marsh.

Out of all this, an understanding developed of much more than fossils. Evolution could be seen in action; that is, the principle of evolution made sense out of observations in science and even everyday life. An obvious case is that of industrial melanism. The peppered moth of England darkened, for better disguise against predators, as trees grew coal-sooty during the Industrial Revolution. In our own lifetimes, with decreasing air pollution, the same species is growing lighter again.

Creationists object that this is not a valid example, but represents mere variability. Nobody, they say, has ever seen a whole new species come into existence. That is true enough, as far as it goes—with some possible exceptions among microscopic organisms. However, evolution takes thousands and millions of years to bring about most of the unmistakable changes that evolutionists describe. The evidence is necessarily indirect. But so, just as necessarily, is the evidence for the reality of events chronicled in the Bible.

Many anomalies have cropped up, but most have turned out to be explainable. Thus, Piltdown man was always an embarrassment, because he did not fit onto any reasonable human evolutionary tree. At last, chemical analysis showed that Piltdown man was a hoax. Without the great guiding principle of evolution in general, who would have paid attention to him at all?

Likewise, we have seen the evolutionary principle in sometimes tragically practical application today, as pathogenic microbes gain immunity to antibiotics through the selfsame process of natural selection that Darwin found. On a still deeper level, we find that we can best understand the details of protein chemistry as between different species (for instance, cytochrome-c) in terms of their differentiation through geological time; but it was the concept of evolution that caused researchers to look for such divergences in the first place.

So we have very briefly reviewed the development of evolutionary thought

—the evolution of evolution, so to speak—and seen how fundamental it has become to biology. Now we must return to the comparison with physics, and to the philosophy of science in general.

As we have seen, theories are subject to disproof. Else they would have no meaning. (Thus, if I told you that space is pervaded by a fluid so subtle that no instrument or experiment can possibly detect it, you could not prove me wrong, but you would not be obliged to take me seriously, either. As a matter of fact, this is precisely what happened to the "luminiferous ether" about which 19th-century physicists had speculated. It turned out to make no difference whether the ether existed or not; therefore nobody had any further reason to imagine that it did exist.)

Thus many theories have fallen by the wayside. But some reveal themselves, in the course of time, to be more fundamental than that. They become basic principles, by which theories themselves are tested. They become touchstones by which observations are evaluated. They become a context within which everything else, in a given field of science, is understandable.

Examples within physics are the two laws of thermodynamics, already mentioned. Without them, we simply could not make sense of our observations of any process involving energy exchange. With them, not only do we comprehend what we see, we are led to new discoveries.

For instance, back in the 1930s, physicists noticed certain curious features of recoil during radioactive decay. The energies and momenta did not balance out as they were supposed to. Either the principles of energy (and momentum) conservation were wrong, or else some ultra-tiny particle was involved, carrying off the excess. Rather than give up their basic principles, which were far too helpful to discard, scientists hypothesized that such a particle did exist: the neutrino. This idea proved fruitful in gaining more knowledge of the nucleus—although not until a generation later was the neutrino actually detected, and then only indirectly.

Granted, basic principles originate in empirical observations. Indeed, the laws of thermodynamics came out of grubby engineering work, and rather late in the history of science at that. Nor are the basic principles Holy Writ. They are subject to modification as our knowledge grows. Thus the separate principles of conservation of mass and energy were unified—modified—into the single principle of the conservation of mass-energy by Einstein.

However, such principles have become so fundamental that the complete overthrow of any of them would mean the complete overthrow of the sciences with which they are concerned. We would be practically back to Square One. It is therefore both understandable and sensible that scientists will not—cannot—set them aside without an absolutely overwhelming, and hence unlikely, body of evidence.

I submit that evolution is no longer a mere theory. It has become just such a basic principle. It is as much a fundamental of the universe, as we conceive the universe to be, as are the laws of thermodynamics or relativity. There is no scientific argument against it, only an antiscientific one.

The question remains: How shall we persuade a lot of perfectly nice people that they are undermining a cornerstone of their entire civilization?

Rick Cook

THE LONG STERN CHASE: A SPECULATIVE EXERCISE

JULY 1986

GIVEN that evolution happens, why did we turn out the way we did? What was there in our ancestors' past that led them to develop the peculiar characteristics that make us "human"? For that matter, what *are* those characteristics? Everybody can rattle off a few obvious traits that seem to set man apart from the other animals, but the author of this piece makes a good case for the idea that certain less obvious, less impressive qualities may be no less important.

Rick Cook has published insightful, thought-provoking articles on a wide variety of subjects and does a great deal of work for computer magazines. He and his wife live in Phoenix, surrounded by cats, books, and computers, "not necessarily in that order." Recently he has extended his writing efforts to fiction, with several short stories well received in *Analog*, one novel, *Wizard's Bane*, out from Baen Books and another, *Limbo System*, on its way. ∎

The Long Stern Chase: A Speculative Exercise / 25

WHERE is everybody?

By rights, there should be hundreds or thousands of intelligent species in this galaxy alone. Our skies and possibly our solar system should be riddled with their spoor. Over just a few thousand years, even one space-going race could be expected to spread its sign through the entire galaxy. Yet there is nothing; not one single verifiable indication of an extraterrestrial species anywhere. Why not?

There are a number of possible answers. Perhaps we know less about the universe than we think we do. Perhaps we are not looking in the right places. Perhaps civilizations collapse within a few thousand years of their birth.

Or perhaps the reason is psychological. Perhaps the universe is indeed full of intelligent races, but for some inner reason none of them, or almost none of them, have chosen to venture beyond their own skies. Perhaps there is some innate difference between humans and Others that drives us to look to the stars where the rest are content to stay home.

We are dealing in moonshine, of course. The entire case for life on other worlds is built on a combination of speculation, a little knowledge, and a firm belief that having just one intelligent species in the galaxy is stretching the odds all out of shape. The astrophysics of life-bearing planets is largely speculative, xenobiology is almost entirely speculative, and as for xenopsychology, well . . .

But just to be contrary, let's suppose that there is a difference in the way our minds work that makes us more likely than most species to go starfaring, either directly or through the agency of our machines. Suppose there is something that fundamentally sets us apart from other animals, something other than intelligence, and suppose that that something has psychological implications.

Just suppose . . .

THE DIFFERENCES

Physically and physiologically, there are a lot of differences between humans and most other animals, and many of them even mark us off from our close cousins, the primates.

Ironically, the most noteworthy differences are not the most noted. When we set ourselves apart from other animals, we tend to do so on the basis of our large brains, erect posture, binocular vision, and mobile hands with opposable thumbs. These are differences, but except for the erect posture, they are all only exaggerations of traits that run through the primate family tree. For that matter, our closest living relatives, chimps and gorillas, go semi-erect part of the time.

But there is a second set of characteristics humans have, ones that we don't seem to notice, which also makes us different. And by and large, they are not shared with other primates.

Perhaps the most obvious one is that humans are almost hairless. Monkeys and apes are hairy, although chimpanzees are more sparsely covered than others. Hairless land mammals are rather rare, and such hairlessness is almost always for a reason. Armadillos and pangolins are hairless because they are covered with armor, for instance. Very large tropical mammals such as elephants and rhinoceri have a lot of bulk for their surface area and a need to radiate excess body heat. Yet humans, who are only medium-large mammals with no armor and just as much need for protection from heat and cold as any other mammals, have little hair.

Mention of our erect posture hints at another obvious difference between humans and apes—our shape. Compared to other primates, humans have very long legs, relatively short arms, and an elaborate musculo-skeletal adaptation for walking upright including very highly specialized foot and ankle structure.

Our feet and ankles are much more highly specialized and atypical of primates than our much-vaunted skulls. What's more, while the differences reflected in our skulls appeared late and slowly in human evolution, the earliest known hominids had feet and legs that looked like smaller versions of our own.

> *"But could he walk upright?" I persisted.*
> *"My friend, he could walk upright. Explain to him what a hamburger was and he could beat you to the nearest McDonald's nine times out of ten."*
>
> —Paleoanthropologists Donald Johanson and Owen Lovejoy, discussing the oldest known hominid.

Although this is obviously related to our ground-dwelling habits, ground dwelling alone is not enough to explain it. Baboons, which are perhaps ecologically closest of all primates to the African hominids, move almost exclusively on all fours. Alone among the primates, the species of the human line are relentlessly upright, and we have been so for at least three or four million years.

There is another difference between humans and other mammals which is less obvious, but quite striking once you stop and think about it: Humans sweat.

Granny used to say that only horses sweat. Men perspire, she admonished, and ladies glow. Granny was wrong. Men sweat. Women sweat. Comparatively, horses only glow. A sweating horse is a horse under stress and a horse in need of special care. It must be walked and cooled down, and it cannot be allowed to drink all the water it wants. A sweating, exhausted human only needs to sit down and cool off. Actually, a sweating human doesn't even need that. As long as he or she can replenish the water and electrolytes lost in the sweat, a sweating human is perfectly all right.

Sweating from the skin is not common in mammals, and horses sweat more than most. Dogs, cats, and most other mammals don't sweat at all over most of their bodies. Instead, they pant to dump excess heat and supplement that with sweat glands in the paw pads. Even elephants, which are hairless and need to get rid of a lot of heat, don't sweat much. Instead, they wet themselves down from a stream or water hole. Humans have more sweat glands per square inch of skin than any other animal, and we sweat profusely.

One result is that humans can survive at temperatures which would cook a steak. To a human 20 minutes or so in a sauna bath is only slightly stressful. But never take a dog into a sauna. It is likely to suffer heat stroke.

There is one other important point about these differences between Man and Monkey. They are not a laundry list; they are a system. What we have here is an elaborate set of interrelated characteristics. They reinforce each other in adapting humans for . . . what?

THE ADAPTATION

Hairlessness, long legs, copious sweating. All of these are relatively rare characteristics. Put them together and you have a combination that is absolutely unique.

Again, these are not random or recent characteristics. The earliest proto-human fossils from Africa show surprisingly modern leg and ankle development. The other traits are impossible to trace in the fossil record, but as far as we can determine, they seem to be equally ancient.

Do these traits point to any kind of adaptation on the part of early humans? Is there a kind of life that requires long legs, hairlessness, and sweating?

The obvious answer is running. The conventional explanation is that humans are adapted to run to escape predators in a plain/savannah environment, the kind that prevailed in Africa a few million years ago.

Well, yes. Humans are obviously adapted to running. But there are serious objections to the notion that we are adapted to running as a defense mechanism.

The biggest objection is that humans are slow. Over a distance of a few hundred yards, most of the running mammals of about human body weight can leave a human in the dust. Not even an Olympic sprinter can keep up with the average antelope or deer. Very much more to the point, almost all of the springing/running predators such as cheetahs can outrun an Olympic champion over that distance too.

A secondary objection is that other running animals, either predators or prey, don't show the same adaptations. They are not, for example, hairless. Nor do they sweat the way humans do.

And yet we are clearly runners. The structure of the foot and ankle, and the length of our legs all make that obvious. But we seem to be runners of a very different sort. Is there an adaptation which would require the entire pattern of differences between humans and their apelike ancestors?

As it happens there is: an adaptation that is still practiced by humans in some parts of the world today. Humans are perhaps the finest cursorial hunters that have ever existed on this planet.

A cursorial hunter gets game by running it down. They aren't sprinters, like cheetahs; they aren't stalkers and pouncers like tigers and other big cats. They simply pick their target and run it into the ground.

Cursorial hunters aren't common. Among land animals about the only examples are humans and some canines, such as wolves. However, canines usually aren't as relentless in their hunting as humans. They prefer to drift along with a herd of animals and attack opportunistically.

> *A lion wakes up each morning thinking, "All I've got to do today is run faster than the slowest antelope."*
>
> *An antelope wakes up thinking, "All I've got to do today is run faster than the fastest lion."*
>
> *A human wakes up thinking, "To hell with who's fastest, I'll outlast the bastards."*

Humans who hunt cursorially will force the pace. That is the basis of their whole hunting strategy. Also unlike wolves, they are as likely to pick a healthy animal as a weak one. Typically, these modern hunters start by trying to sneak up on the animal and nail it with a thrown spear. If that works, fine, they're spared the exercise. If it doesn't work, the animal skitters off and the hunters follow. After a few hundred yards or a mile or so, the animal tries to settle back down to grazing. But the humans trot up and try again. As the animal runs, the humans stay with it, never giving it a chance to rest. They chase it, and they keep chasing it until the animal can be brought down. Eventually they either get it with a lucky shot or they get the animal when it is exhausted.

Cursorial hunting puts some major demands on its practitioners. A cursorial hunter doesn't have to be much of a sprinter, but it does need enormous reserves of endurance. It must be capable of hours of sustained effort. That means it needs a rather different set of adaptations from a sprinting hunter.

A sprinter must be fast off the mark but it doesn't need a lot of endurance. The issue will be decided in less than two minutes, so the animal's endurance simply doesn't come into play. Efficiency of locomotion is much less important than the ability to deliver a burst of speed.

There is another thing a sprinter doesn't have to worry about: getting rid of excess heat. A sprinting hunter simply isn't in action long enough to build up a lot of heat from exertion. On the other hand, a cursorial hunter is in action for hours. It builds up enormous amounts of heat, and it must have an efficient way of getting rid of it. A cursorial hunter needs an outstanding cardiovascular system and a very good method of dumping waste heat. Humans have these qualities in spades.

The combination of hairlessness and a lot of sweat glands gives us a very efficient way of getting rid of heat produced by prolonged exercise. Humans can stay active in combinations of temperature and humidity that would fell most other animals. "Only mad dogs and Englishmen go out in the noonday sun," but only the Englishmen are likely to last.

As cursorialists, humans have another problem, one not shared by the wolves and other canines. Canines, felines, and most of the larger prey species such as antelopes, sheep, and bovids have a special cooling system built into their skulls to keep their brain temperatures down under strenuous exercise. It consists of a net of blood vessels around the brain that removes excess heat and dumps it into the nostrils. One of the reasons that canines have long snouts is that they have a heat exchanger in there. Primates never developed that extra cooling mechanism, so we have to dump heat from our whole bodies to keep our brains from cooking into heat stroke.

Our legs may not make us great sprinters, but they can carry us efficiently over long distances. A quadruped uses its front legs primarily as shock absorbers while the thrust comes from the rear legs. This means quadrupeds are fast, but each time the front legs hit, energy is lost. Human feet and ankles are adapted differently. The built-in spring mechanism of muscles and tendons in our feet lets us recover and transfer much of the energy expended every time we take a step.

Over long distances, bipedalism is one of the most energy-efficient forms of locomotion known. Perhaps the only more efficient way is bipedal hopping, like kangaroos. But there are trade-offs to consider. Bipeds are not as fast off the mark or maneuverable as quadrupeds, and hopping bipeds are much less maneuverable than running bipeds. If a kangaroo tried to run down an antelope, the antelope could escape by a display of broken field work. Against a running biped, dodging and weaving is a lot less effective.

Aside from the wolf, the only land animal that comes close to humans in long-range endurance is the horse—an animal which has been bred for thousands of years for speed and endurance. Yet humans can run down horses.

When the Apaches raided through Arizona, New Mexico, and Texas a century ago, they raided on foot. They stole a lot of horses, but mostly they ate them. Apache war parties could and frequently did outdistance mounted pursuers.

The physical effects of our heritage are so obvious that we take them for granted. Every year hundreds of thousands of people run 26-mile foot races, and thousands of them finish in times that would kill an antelope or cheetah. In less developed societies, people routinely run even longer distances. Southwestern Indians have a tradition of footraces of 40 to about 100 miles. We honor competitors in these events, but we don't see them as super-human. Spending your vacation hiking 200 miles may make good cocktail party conversation, but it is barely noteworthy to the world at large.

BIOLOGICAL IMPLICATIONS

Cursorial hunters are not common. Aside from man and some canines, there are almost none of them today, and there is no sign there were ever very many of them. As a result, the package of physiological adaptations that goes with that niche is rare, too.

One of the reasons cursorialism is rare is that it is not terribly energy efficient. A cursorial hunter must invest a tremendous amount of time and energy in making a single kill. If the hunter loses the prey in mid-chase, the energy expenditure will be quite large, and for any hunter, failure is always more likely than success.

By contrast, a stalking hunter has very little energy investment in each kill.

It moves in on its prey by easy stages and makes only a short final rush. Even a sprinting hunter like a cheetah has less of an energy investment in each kill than a cursorial hunter, who must make a series of long chases.

About the only time cursorialism is clearly superior is when the available prey is large, but fairly rare. A cursorial hunter has more chases than kills, but its batting average will be much higher than a sprinting or stalking hunter's. A sprinter or stalker essentially only gets one chance at a prey animal. A cursorialist gets another opportunity each time it gets close enough. Unless the prey is able to lose the cursorialist somehow, it is doomed. If the prey species are large enough to feed the hunters for several days, then cursorialism may have an ecological advantage.

We know that in the time and place where the human adaptations evolved, there were prey species of the appropriate size. But there seem to have been too many members of those species and too many other sources of food for cursorialism to have been really energy efficient.

Because cursorialism is relatively inefficient, cursorial hunters typically supplement their diet in other ways. Wolves and other canines get a great deal of nourishment from mice, rabbits, and such, which they do not hunt cursorially. Humans, of course, have been scroungers since the very beginning. In spite of our adaptations, our ancestors probably got at least as much nourishment from gathering plant foods, insects, and small animals as they did from hunting down large game.

The idea that protohumans got a lot of food by gathering and scavenging is confirmed by studies of known protohuman sites and protohuman teeth. The teeth are worn from gritty vegetable matter, and the bones found at the camp sites may have come from scavenged carcasses rather than kills.

Teeth, by the way, are the one part of human anatomy and physiology which doesn't fit this pattern of a cursorial hunter. Our teeth, and the teeth of our known hominid ancestors, are not terribly well adapted to the life of a hunter. A human or any known protohuman who tried to kill anything bigger than a rabbit with its teeth would stand an excellent chance of getting them kicked down its throat.

One possible explanation for this mismatch is that human teeth never had to do the job of a typical carnivore's. A carnivore uses its teeth not only for eating, but also for killing its prey. It needs fangs to hold and slash as well as teeth to slice off meat. Protohumans may very well have done their killing by throwing stones or sticks at their prey from close range or clubbing it to death. Some monkeys will throw things in defense, and chimpanzees have been seen using branches as clubs. Protohumans may have found that it made more sense to beat their food to death than to jump it.

Tool using in protohumans appears quite early. The earliest known human habitation and kill sites, at least two million years old, show large quantities of crude stone tools.

Something else interesting shows up at those early camp sites as well: food sharing. It appears (on admittedly weak evidence) that protohumans brought their food back to camp to consume with their fellows. This is extremely unusual behavior for primates. Many monkeys and apes forage in groups, but except in the case of mothers and infants, they don't share what they find. The closest they come is in chimpanzees which practice "tolerated scrounging" when meat becomes available. Even the early ancestors of man, on the other hand, apparently brought food back to camp.

Not even most carnivores do this. They may drag their kills to a more secluded spot, although usually the carnivores that live in groups, like lions, don't even do that. Canines, our fellow cursorialists, come closest to that pattern.

In the context of cursorialism, bringing food back to camp to share makes excellent sense. If you have to chase your prey for miles, it is more energy efficient to bring it back to a central point than to have everyone come to the kill site.

This is particularly true if you have divided your forces. If some of the group members are out hunting and others are foraging for plants and small animals, you have a much better chance of not going hungry even if you don't get the high-quality protein food you wanted. Since females are saddled with children, they are the logical ones to do the gathering—which may explain the beginnings of human society.

THE SOCIOBIOLOGY OF CURSORIALISM

Assuming that Man did indeed evolve as a cursorial hunter, what implications might this have for the way we view the universe?

The notion that biology is destiny is nonsense. But very few would deny that biology influences the way we behave and look at the world. Poul Anderson once said that while technology does not completely determine the structure and goals of a society, it definitely sets the possibilities. At the very least, the same thing can be said of our biological heritage.

Our physiology has unquestionably influenced our ways of thinking and perceiving the world in other areas, so it seems reasonable to assume that our adaptation to cursorial hunting had psychological consequences which linger to this day.

That being the case, what psychological traits would cursorialism be likely to emphasize?

The most important one is persistence. A cursorial hunter must keep after the prey, press it closely, and never give it the opportunity to rest. "A stern chase is a long chase" was a saying well known to captains in the days of sailing navies. By its nature a cursorial hunt is usually a stern chase—one the hunter cannot afford to give up.

> "If at first you don't succeed . . . To Hell with it."
>
> —Tony the Tiger/Leo The Lion/Charlie the Cheetah

This is very different from most other hunters. A big cat will make its charge and veer off if the prey escapes, and is no more likely to go after that animal again than any other animal in the herd. We call persistence "doggedness" in honor of our ancient hunting companion, but even most canines are unreliable trailers. After thousands of years of selective breeding, most of them are still easily distracted if something more interesting crosses their path.

When we speak of a cat "crouching for hours" beside a mousehole, we are indulging in anthropomorphism. A close examination will show that the cat spends most of those hours by the mousehole asleep. (Cats sleep about 18 hours a day.)

A human, on the other hand, will spend hours crouching beside a figurative mousehole. Or days or weeks or years if the "mouse" is big enough. In fact, if the mouse is really big, we will dedicate our lives to watching that mousehole and set our children and grandchildren to watching it as well.

This kind of persistence is such a human characteristic that one of the things we measure as part of intelligence is persistence. We assume that humans are more persistent than animals because we are more intelligent. To us, someone with a "short attention span" is unintelligent or childish. Adult humans in all cultures and at all levels of development persist.

Yet how much evidence do we really have that perseverance is an innate part of intelligence? Even in humans, persistence seems to vary independently of intelligence. In our own culture there are a lot of bright, flighty people: people who are obviously intelligent but who can't stick to a subject in conversation, to an idea, or to a task. Conversely, we recognize that persistence can substitute for intelligence in many situations.

Persistence in this sense means pursuing some goal (notice how the language reflects the chase?) rather than mechanically repeating a series of actions. Ants will spend days trying to climb an obstacle between them and the nest. Rats and other animals (including humans) can be conditioned to do something endlessly. In repetitive behavior like this, however, the behavior is the thing. The ant keeps trying to climb the obstacle because the scent trail takes it there. A conditioned animal will continue the behavior no matter how inappropriate. In persistent behavior (to create a distinction), the goal and not the action is the thing. If the antelope curves off to the right, try to get inside him rather than blindly following in its tracks. If X doesn't work, try Y and if that fails, there is always Z, and beyond that try A again.

Suppose, then, that persistence is not a necessary concomitant of intelligence. Suppose it is merely another characteristic humans happen to have in addition to intelligence. And suppose that most of the other intelligent races of the galaxy are not descended from cursorial hunters and aren't nearly as persistent as we are.

THE IMPLICATIONS FOR THE UNIVERSE

As we have noted, cursorial hunters are relatively rare. There is no reason to think they would be any more common in other ecologies or on other planets. If we assume that high persistence correlates with cursorialism, it would appear that few or no other intelligent races would have as much of it as humans do. So what then?

Less persistent races probably aren't less likely to develop civilizations. Civilization seems to arise in response to a need to control the external environment, and that will exist on most planets with intelligent life. Less persistent species probably aren't any less likely to develop technological civilization, since that represents an extremely efficient adaptation to the environment (at least in the short run). Because persistence has been so important in our scientific advancement, it might take them longer than it has taken us, but there is no reason to think that they would not ultimately develop technological civilization.

Assuming that their solar systems are like our own, most of those technological civilizations will quite likely develop interplanetary travel sooner or later. A technological civilization will probably expand to the limits of its resources, and all of a solar system represents resources available for the taking.

But would a low-persistence civilization go beyond interplanetary flight?

Probably not. Barring some kind of yet-undiscovered faster than light drive which is easy to find and easy to build, the effort it would take would simply be too great for the rewards. Humans, with their heritage of seeking long-distance goals, might be willing to build generation ships, or send self-replicating robots to the stars, but for a race without human persistence, it might well be too much work for too little result. Even the highly technological civilizations would be content to stay in their own systems. Stray radio transmissions would be detectable in their immediate neighborhood, but those fade out over the distance of a few light years.

Such beings might well be smarter than we are, and they might be much more capable in the short term, but without our inbred doggedness, they would not become starfarers.

Absent some magical, serendipitous discovery, the road out of any solar system would be long and hard. It would have to be a goal for generations, with many setbacks and disappointments along the way.

Going to the stars would be the longest, hardest stern chase of all—and species not bred to such a chase could well find it beyond them.

Author Note The Johanson-Lovejoy quote is from *Lucy: The Beginnings of Human kind* by Johanson and Edy (Simon and Schuster, 1981).

The idea of hominids as cursorial hunters was suggested to me by Bill Vaughan.

Christopher P. Dunn

ADVANCED

MACHINING IN

ANCIENT EGYPT?

AUGUST 1984

SOME of our ancestors left remarkable evidence of their passing, such as the great pyramids of Egypt, but not enough details for us to understand exactly how they did these things. Naturally, such objects have stimulated all manner of speculations, up to and including the idea that their makers must have had extraterrestrial help. Such extreme hypotheses, supported by little or no hard data, may not be necessary and, in fact, may do our ancestors an injustice: They may have known how to do far more on their own than we commonly give them credit for. This article, for example, takes a very close look at some superficially mundane objects and concludes that the ancient Egyptians just may have had some surprisingly modern manufacturing techniques.

It's hardly surprising that Christopher P. Dunn should look at these relics of the past from the unusual perspective of an industrialist's eye. He has 28 years of personal experience in manufacturing, from a youthful apprenticeship in Manchester, England, through stints as machinist, toolmaker, process engineer, laser applications engineer, and laser job-shop manager in the United States. He is presently an engineer with a midwestern aerospace manufacturing company. His interest in ancient Egypt was piqued in 1977 when someone "dared to suggest that the Great Pyramid wasn't a tomb." ∎

Advanced Machining in Ancient Egypt? / 35

THE popular topic of ancient civilizations—and speculation that some cultures were perhaps more advanced than what is generally believed—has frequently held our fascination and, in varying degrees, our belief. Velikovsky, Pauwels and Bergier, and Von Däniken have attempted to persuade the world that a huge misconception exists regarding our study and beliefs of pre-history. In most cases, however, the introduction of new ideas leaves more unanswered questions than answered ones.

Fragments of evidence uncovered here and there seem to suggest that certain artifacts were created for technological purposes by technologically advanced people. Ancient maps which depict Antarctica without its ice cap; strange lines on the Nazca Plains in Peru; prehistoric earthworks in Ross County, Ohio; iron pillars which have stood for thousands of years without rusting; the Pyramids of Giza; and countless other artifacts might appear to have been created by people who had more than religion and war on their minds.

While interest in prehistoric anomalies simmered along for many years, the subject exploded into a furor of debate when Von Däniken invoked the name of God (*Chariots of the Gods?*, 1969, G.P. Putnam's Sons) and proceeded to sell millions of books with that name boldly emblazoned on the cover. His attempt to persuade the world that space travelers had influenced early civilizations did not impress most historians and archeologists, who perceived early man as practicing a less exotic profession. But even in his fundamental endeavors, do we fully understand the degree of sophistication that prehistoric man had attained?

Archeology is largely the study of man the toolmaker; for it is with tools, and artifacts created with tools, that an understanding of past civilizations can evolve. For the most part, the tools that have been uncovered are adequate in explaining artifacts of the same period. However, there seems to be a time in history where a glaring discrepancy exists between the artifacts that were discovered and the tools which supposedly created them.

The ancient Egyptians left artifacts behind that cannot be explained in simple terms. Supposedly contemporaneous tools have been uncovered which do not fully represent the "state-of-the-art" that is physically evident in some of the most intriguing objects available from a civilization which, despite its most visible and impressive monuments, has left us with only a sketchy understanding of its full experience on planet Earth.

Sir William Flinders Petrie left some unfinished business when he wrote *Pyramids and Temples of Gizeh* in 1883. After closely examining several artifacts (Figure 1) he had found on an expedition to Egypt (undertaken in 1880), he was left with questions he could not answer, and had to draw his investigation to a close, resigned to the fact that the ancient Egyptians were able to cut granite—a hard igneous rock—using methods that were not only efficient, but also quite remarkable. Regarding a hole that was drilled into a piece of granite he wrote, "the spiral of the cut sinks .100 inch in the circumference of 6 inches, or 1 in 60, a rate of ploughing out of the quartz and felspar which is astonishing."

The characteristics of some of his finds suggest methods of cutting granite

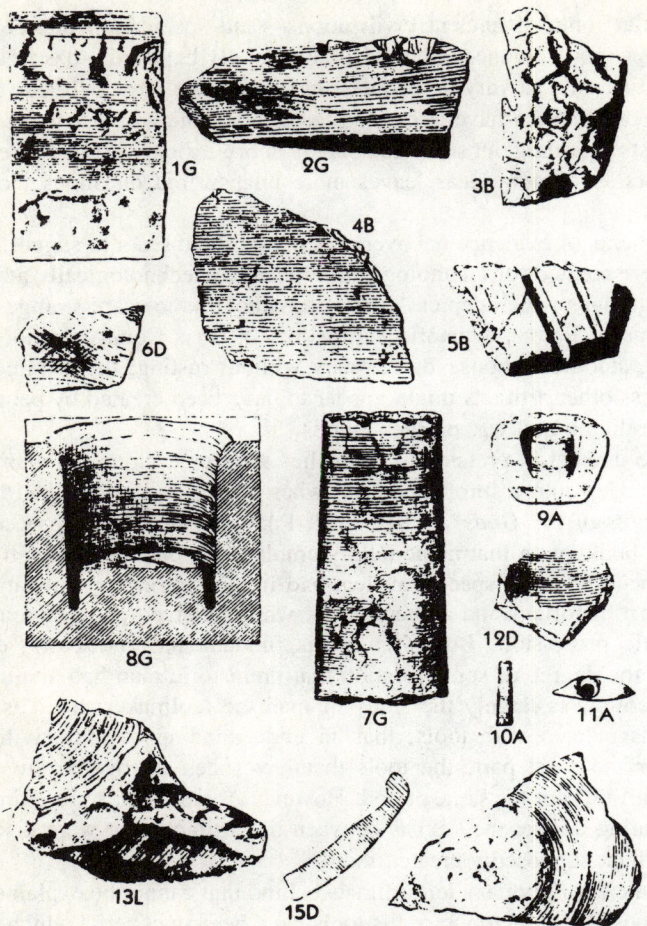

FIGURE 1. A—Alabaster B—Basalt D—Diorite G—Granite L—Limestone —From Pyramids and Temples of Gizeh, by William Flinders Petrie, copyright 1883, London. Reprinted by courtesy of Ann F. Petrie. A reprint edition of the book is planned by Akadem. Druck-u. Verlagsanstalt, Graz, Austria.

and diorite that were unknown in his time; and though Petrie attempted to explain the technical principles employed in cutting the rock, they displayed subtle characteristics he could not discern. It is these details which stand them apart from other relics of the past!

Petrie reached a conclusion that the ancient masons used bronze saws tipped with jewels to cut the huge monoliths that were found on the Giza Plateau. Exceeding eight feet in length, according to Petrie these saws were instrumental in building the most monumental work ever undertaken by any civilization the world has sustained.

Nothing has been found of these saws or other equipment that must have been used to cut and erect the pyramids and temples of Egypt. Some may point to the few copper chisels that were found near the pyramids. But chisels are inadequate in explaining the machining marks described by Petrie, and cannot be seriously taken into consideration when faced with hard igneous rock—which the Egyptians were able to cut with such ease.

Petrie presents evidence that would certainly indicate the use of saws on this granite, but what kind of saws they were has not been fully described. So perhaps, a hundred years later, we should take another look at Petrie's work and re-examine the methods of the ancient Egyptian stonemasons.

In this paper it is assumed that Petrie's observations were accurate—but we must recognize that we are dealing with secondary evidence. However, as Petrie was not attempting to proselytize and was only objectively reporting the data he had uncovered while searching for answers, we can be reasonably sure that it was not invented, and that it should stand unless proved wrong.

Alternative conclusions offered here are no doubt controversial; nevertheless, I believe them to be logical. They follow a process of elimination in a search for the truth in what is a most unusual and intriguing facet of Egyptian craftsmanship.

Methods employed to produce ancient artifacts can usually be explained in terms of simplicity. The hammer was probably the first tool invented, and hammer-working gold, silver, and copper is known to have produced some of the most beautiful and intricate pieces of work in existence today. Since early man first realized that he could effect changes in his environment simply by applying force with a reasonable degree of accuracy, the development of tools has been a continuous and fascinating part of the growth of civilization.

It is ironic that the basic material that started man on his quest for technology is also the one that has survived to tell us the degree of progress the ancient Egyptians had made 50 centuries ago.

The millions of tons of rock that the Egyptians had quarried for their pyramids and temples—and cut with such superb accuracy—reveal glimpses of a civilization that was technically more advanced than what is generally believed. Even though it is thought that millions of tons of rock were cut with simple primitive hand-tools, such as copper chisels, adzes and wooden mallets, substantial evidence shows that this is simply not the case. Even discounting the argument that work-hardened copper would not be suitable for cutting igneous rock, other evidence forces us to look a little harder, and more objectively, when explaining the manufacturing marks scoured on ancient granite by ancient stone craftsmen.

Who doesn't recognize the resources—both material and human—that must have gone into building the Egyptian pyramids and temples? We know that thousands of tons of granite were cut and transported over a distance of 500 miles from the Aswan Quarry—this particular granite is a reddish color, with 55 percent silicon quartz crystal. A substantial amount found a home deep in the heart of the Great

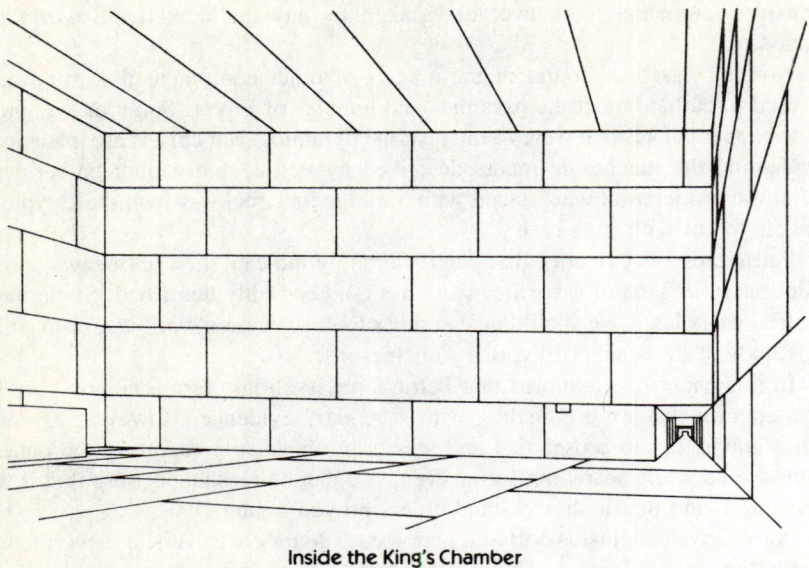

Inside the King's Chamber

FIGURE 2. The weight of speculation surrounding this chamber may exceed the weight of the stone. Though no proof of a burial has been found, the King's Chamber has managed to keep its name.

Pyramid where an entire complex was built—later to become known as the King's Chamber (Figure 2). Included in this complex were 43 giant granite beams weighing between 45 and 70 tons each (Figure 3). The feat of transporting and erecting these monoliths has astounded many people, and the methods that were used in the process have been strongly debated. The concern here, though, is not so much how they were carried, but how they were cut. This is more easily explained, mainly because there is more evidence. Let's take a look at this evidence.

Although the Egyptians are not given credit for having a simple wheel, the machine marks they left on the granite found at Giza suggest a much higher degree of technological accomplishment. Petrie's conclusion regarding their mechanical abilities shows a proficiency with the straight saw, circular saw, tube-drill and, surprisingly, even the lathe—the father of all machine tools in existence today. The evidence he presents is quite striking, especially the evidence of lathe-turning techniques, which were evidently of a higher standard than we would expect from an undeveloped industry.

There were two pieces of diorite in Petrie's collection which he claimed must have been the result of true turning on a lathe. The traditional theory of how primitive people shaped hard stone does not apply in the case of these diorite relics. It is true that many intricate objects can be created without the aid of machinery, simply

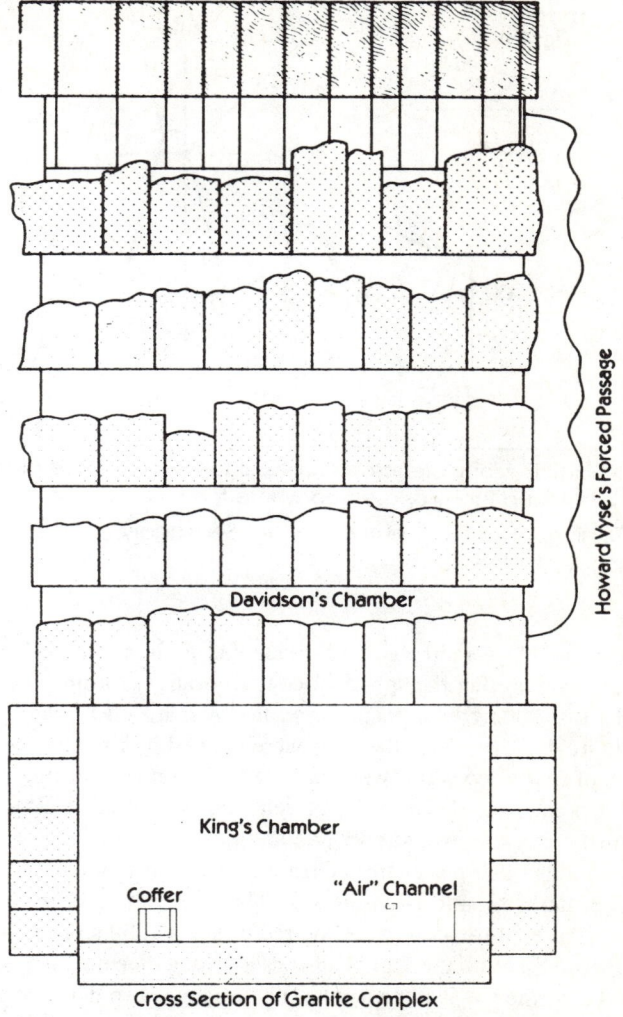

FIGURE 3. A difficult stone to quarry and cut; yet there seems to be a redundancy of granite above the King's Chamber. Above one layer of granite beams, forming the ceiling of the chamber, were hidden four more layers—until Col. Howard Vyse's crew blasted their way through the core masonry during their expedition in 1873.

by rubbing the material with an abrasive, such as sand, using a piece of bone or wood to apply pressure. The relics Petrie was looking at, however, in his words "could not be produced by any grinding or rubbing process which pressed on the surface."

To the inexperienced eye, the objects Petrie was studying would hardly be

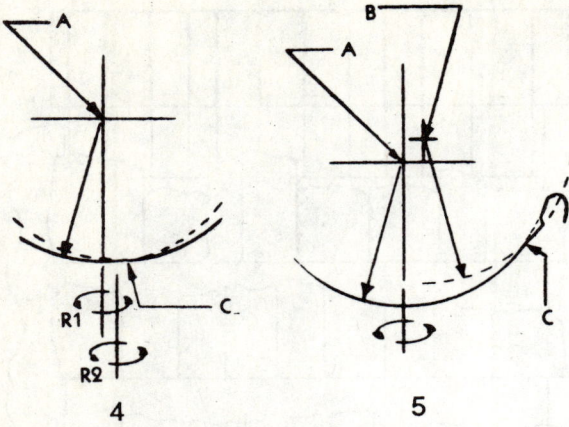

FIGURE 4. R-1 and R-2 indicate rotation axes of the bowl. A is the pivot point of the tool and C is the cusp that was created where the radii intersect.

FIGURE 5. Original pivot point of the tool -A-. Secondary pivot point -B- creating lip and cusp -C- at intersection.

considered remarkable—a simple bowl, made out of simple rock. But on studying the bowl, Petrie found that the spherical concave radius, forming the dish, had an unusual feel to it. Closer examination revealed a sharp cusp where two radii intersected (Figure 4). Evidently, the radii were created by rotating the bowl on two separate axes of rotation when it was machined. How many machinists today have chucked on a premachined part and not centered it correctly? Together with the sharply defined cusp, this would preclude any possibility of hand work.

On examining other pieces from Giza, Petrie found another bowl shard which, again, had the marks of true lathe-turning. This time, though, instead of shifting the workpiece's axis of rotation, a second radius was cut by shifting the pivot point of the tool (Figure 5). With the same radius used to machine the dish, they machined just short of the perimeter, leaving a small lip. Again a cusp defines the intersection of the radii. Obviously, if this had been the result of hand work, the cusp would have been rubbed away—if it had existed in the first place.

Even though the evidence of prehistoric lathe work seems quite conclusive, it doesn't necessarily mean that their lathes were developed to the same degree that they are today. We have assumed that manpower has been the sole motivating force for work produced at that time and, unless proved otherwise, we could logically assume that the ancient masons had developed a primitive form of lathe which utilized manpower, or animal power, though it is remarkable that they were able to cut the rock so cleanly, with such sharply defined features, without splintering.

The transition from manpower to machine power has greatly improved efficiency. When manufacturing modern day artifacts, we seldom pick up a hand saw if a mechanized saw is at hand. When cutting large surface areas, a hand saw is not even thought of. Speed and accuracy are essential ingredients for any job undertaken for profit in today's society. Let's face it, leisurely production levels are a thing of the past—or are they? I believe that the ancient Egyptians might argue with us on this one.

There are subtle characteristics on a granite sarcophagus found in the Great Pyramid, and also on one in the Second Pyramid, which suggest that they were not cut with back-breaking labor by men bent over their crude primitive tools, but with speed and accuracy. . . . Well, almost accuracy.

It is well known that the casing stones on the Great Pyramid were cut and positioned with great care and precision, achieving flatness and parallelism within .010 inch. The granite boxes, on the other hand, were blatantly inferior in workmanship. Out of sight, out of mind, sure; but wouldn't Khufu have had something to say about it, if, of course, the granite box was indeed intended to be his final resting place? The Great Pyramid's box shows visible saw marks on the outside where at two places the saw had cut too deep into the side. The masons had backed the saw out and proceeded to cut again, until two inches lower down where they had to make another restart because the saw had wandered again. The amount of error in this case was .100 inch.

The granite box inside the Second Pyramid is polished all over except for the bottom. In this case, an adjustment was made to correct a .200 inch error. Unlike the box inside the Great Pyramid, this one didn't have visible saw lines on the sides.

Assuming that hand sawing a block of granite 90 inches by 38.68 inches by 41.23 inches would be laboriously slow, it is significant that these errors were made. Can you imagine a team of masons operating a 9-foot-long hand saw zipping through a block of granite so fast that they cut past their guideline .100 to .200 inch before making a correction? Then, to back the saw out, restart their cut on the right line and proceed to make the same mistake 2 inches farther down does nothing to confirm the speculation that this was the result of handwork.

Hand sawing is a slow process at the best of times, and a careful craftsman can see the direction his saw is taking and make corrections before a serious mistake can be made. Manually operated saws can wander like mechanized saws, probably more so considering the increased risk of human error. These considerations, along with nuances inherent in machining practices that experienced craftsmen will recognize, indicate that the ancient stonemasons were using a saw that was cutting through the granite at a faster pace than hand sawing could achieve.

Going from machining the outside of the granite boxes to hollowing out the insides, the stonemasons again made mistakes, only this time with drills. The methods used by the pyramid builders to hollow out the insides of these boxes are the same as methods that would be used today to machine out any component.

Tool marks on the inside of one box indicate that when the granite was hollowed out, preliminary roughing cuts were made by drilling holes around the area that was to be removed. According to Petrie, these holes were made with tube-type drills that left a central core, which was knocked out after the hole was cut. After all the holes had been drilled and the cores removed, the insides were worked by hand to their desired dimensions.

However, the machinists once again let their tools get the better of them, and Petrie found the resulting errors on the inside of the box in the King's Chamber. Their drill had evidently worked its way into the wall of the box, and they were unable to polish out the error without drastically changing the inside dimensions. Even after polishing away $\frac{1}{10}$ inch, they were still left with an error $\frac{1}{10}$ inch deep, 3 inches long, and 1.3 inches wide, located 8 to 9 inches below the top of the box. Two such errors were made, one to a lesser extent.

These mistakes would not be unusual in a modern machine-shop. A long drill would be especially susceptible to "wander" from the perpendicular. One remarkable detail is that they chose this method to hollow out their granite box. The inside depth is just under 3 feet, which would call for a good-sized drill. Regarding this detail we might assume that they had reached a high degree of efficiency with this method of machining.

There are several unknowns, principally their means of detecting the error and the way they managed to correct the drill's course. Other questions arise, such as the removal of waste material from the drill-hole—flooding with water might be one solution—and the speed of their operation. In answer to this last question, we will look at artifacts that were found around Giza and in the Valley Temple. These relics provide the answer to most questions regarding the level of technological progress the ancient Egyptians had made in cutting stone.

While the evidence grows to support the premise that the ancient Egyptians had used some kind of high-speed machinery when cutting the granite found at Giza, we are still left with the question of how this machinery was powered. And while we do not generally think in terms of the ancient Egyptians having discovered and utilized electricity, the following evidence suggests that electricity should not be ruled out when we seek a complete understanding of this civilization's technological accomplishments.

The machine marks left in the tube-drilled holes found in the Valley Temple and the cores that came out of these holes—and numerous others—are without doubt the clearest and strongest argument yet found for considering the ancient pyramid builders as being technologically advanced. Another close look at the characteristics of the granite hole reveals more about the Egyptians than what Petrie was first able to discern.

When searching for a method that fully explains all relevant data without leaving unanswered questions, we have to begin with the simplest possible method. If that does not satisfy the evidence, we must then move on to another method regardless of the direction it may take us. The marks left in the granite hole and core indicate that whatever had cut them had removed material at a rate of .100 inch per revolution of the drill. This is a phenomenal feed rate for drilling into a solid piece of material and certainly not one that could be achieved by hand. (If anybody knows of a way to achieve such a feed rate using conventional machining methods there are several machine tool manufacturers waiting to hear from you.)

Petrie was so astounded by these artifacts that he attempted to explain them at three different points in one chapter. Let's look at what puzzled him so much.

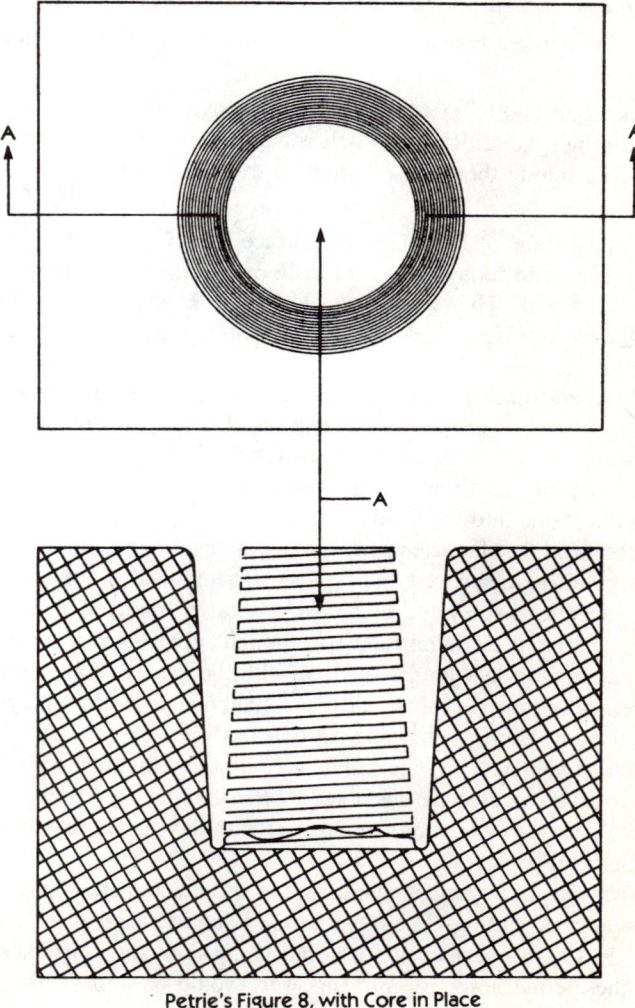

Petrie's Figure 8, with Core in Place

FIGURE 6. In drilling a hole, .100 inch per revolution of a drill is a phenomenal, if not impossible, feed rate for any material. Such a hole was cut into solid granite by ancient Egyptian stonemasons, whose technical skills seem to have been severely underrated in the past.

Again, we have an artifact that could be passed off as insignificant, except that in this case the machine marks were atypical of lathes, or of any other machine tools in existence in Petrie's time. Close examination of the evidence revealed three characteristics of the hole and core that make these artifacts extremely remarkable:

1. A taper on both the hole and the core.

2. A symmetrical helical groove following these tapers and cut at .100 inch per revolution.

3. The confounding fact that the spiral groove cut deeper through the quartz than through the feldspar—which is a softer material. Logically, in conventional machining the reverse would be true (Figure 6).

Mr. Donald Rahn of Rahn Granite Surface Plate Co., Dayton, Ohio, tells me that in drilling granite today, diamond drills penetrate at the rate of 1 inch in 5 minutes at 900 R.P.M. This works out at just .0002 inch per revolution, which means that the ancient Egyptians were able to cut their granite with a feed rate 500 times faster.

But there are other problems. They cut a tapered hole, with grooves, and these grooves were cut deeper through the harder constituent of the granite. If conventional machining methods can't answer just one of these problems, where do we look to answer all three? I hope that you are as baffled as I was when this information first came into my hands. Think! Is there a primitive method of cutting that would create all the characteristics of the granite artifacts? If there is, the answer that follows could be disproved. (A typical archeological process is to reproduce artifacts in order to display the technical processes involved in creating them.) Contact with the granite-cutting industry, though, gives us reason to consider the following. Perhaps those who are familiar with the subject may want to ponder a while before reading the rest of this article. Challenge yourself! Can you determine the method used by the Egyptians to cut holes in granite?

In the final analysis, there has to be no doubt that the method being described could achieve all the characteristics noted by Petrie. And keep in mind the methods proposed by Petrie—that is, a bronze tube set with jewels and sustaining a pressure of 2 to 3 tons as it revolves.

The fact that the spiral is symmetrical is quite remarkable considering the proposed method of cutting. The taper indicates an increase in the cutting surface area of the drill as it cut deeper, hence an increase in the resistance. A uniform feed under these conditions, using manpower, would be impossible.

The suggestion that jewels set in bronze could develop the configurations in the granite does not take into consideration the following question: Under several thousand pounds pressure, which material would offer the most resistance, the granite or the bronze? The jewels would undoubtedly work their way into the softer substance, leaving the granite relatively unscathed after the attack. Nor does this method explain the groove's being deeper through the quartz! This is the most compelling piece of evidence to support the mode of cutting that is about to be described.

This method will explain how the holes and cores found at Giza could have been cut. It will create all the details that Petrie and initially I puzzled over so much. Unfortunately for Petrie, the method was not known at the time he made his studies, so it's not surprising that he couldn't find any satisfactory answers. The application of ultrasonics in machining brittle material, while being relatively new to us, could very well be the only method that completely satisfies logic and explains all noted phenomena.

Ultrasonic machining is the oscillatory motion of a tool which chips away material. Like a jackhammer chipping away at a piece of concrete pavement, except much faster and not as measurable in its reciprocation, the ultrasonic tool-bit, vibrating at 19,000 to 25,000 hertz, has found unique applications in the precision machining of odd shaped holes in hard, brittle material such as hardened steels, carbides, ceramics, and semiconductors. An abrasive slurry or paste is used to accelerate the cutting action.

Using the following method, the Egyptians moved more material than they actually machined. In fact, regarding attaining maximum results for the least amount of effort, the method is completely logical.

The most significant detail of the drilled hole is the groove that is cut deeper through the quartz than the feldspar. Quartz crystals are employed in the production of ultrasonic sound and, conversely, are responsive to the influence of vibration in the ultrasonic ranges—i.e., they may be made to vibrate at high frequencies. In machining granite using ultrasonics, the harder material would not necessarily offer more resistance, as it might during conventional machining practices. In fact, an ultrasonically vibrating tool-bit may find numerous sympathetic partners while cutting through granite—embedded in the granite itself! Instead of resisting the cutting action, as it would normally, the quartz could be induced to respond and vibrate in sympathy with the high frequency waves and amplify the abrasive action as the tool cut through it.

The fact that there is a groove may be explained several ways. An uneven flow of energy may have caused the tool to oscillate more on one side than the other. The tool may have been improperly mounted. A buildup of abrasive on one side of the tool may have cut the groove as the tool spiraled into the granite.

That the hole and the core have tapered sides is perfectly normal if we consider the basic requirements for all types of cutting tools. This requirement is that clearance be provided between the tool's non-machining surfaces and the workpiece (Figure 7). Instead of having a straight tube, therefore, we would have a tube with a wall thickness that gradually became thinner along its length. The outside diameter would gradually get smaller, creating clearance between the tool and the hole, and the inside diameter gradually would get bigger, creating clearance between the tool and the central core (Figure 8).

As shown in Figure 8, this would allow a free flow of abrasive slurry to reach the cutting area, and perhaps the waste could be vacuumed out through the central

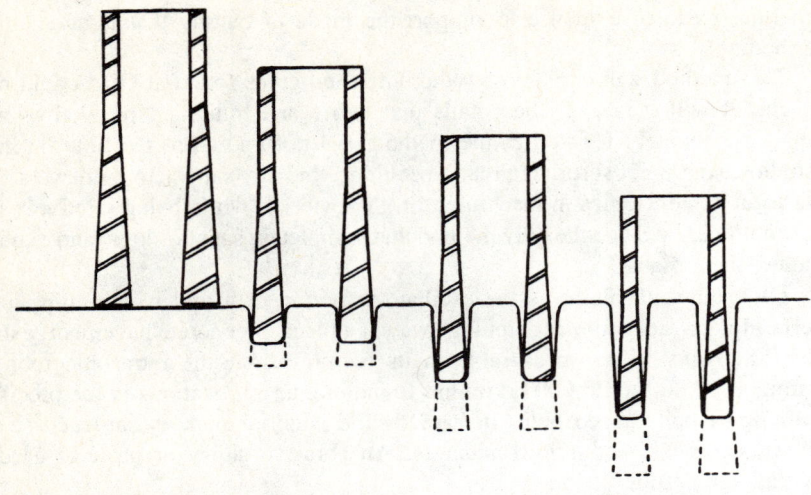

FIGURE 7. As the tube drill sinks into the granite, wear of the cutting surface is constantly changing its dimensions. This creates taper on both the core and the hole. The dotted line illustrates the original tube size.

hole. It would also explain the tapering of the sides of the hole and the core. Since the tube-drill was a softer material than the abrasive, the cutting edge would gradually wear away. The dimension of the hole would correspond to the dimension of the tool at the cutting edge. As the tool became worn, the hole and the core would reflect this wear in the form of a taper.

The spiral groove can be explained if we consider one of the methods that is predominantly used to advance machine components uniformly. Using a screw and nut method, as shown in Figure 8, the tube-drill could be efficiently advanced into the workpiece by turning the handles (*a*) in a clockwise direction. The screw (*b*) would gradually thread out of the nut (*c*), forcing the oscillating drill into the granite. It would be the ultrasonically induced motion of the drill that would do the cutting, and not the rotation—the latter being used purely to sustain a cutting action at the workplace.

The theory of ultrasonic machining may resolve all the unanswered questions that other theories do not. Methods may be proposed that might cover any singular aspect of the machine marks, and not progress to the exotic method described in this paper. It is when we search for a single method that will provide an answer for all collective data—and leave no doubt as to its validity—that we move away from primitive and even conventional means of manufacture and are forced to consider evidence which could settle once and for all the debate regarding the ancient Egyptians' level of technology.

We are still left with some very serious questions regarding their culture. The

Advanced Machining in Ancient Egypt? / 47

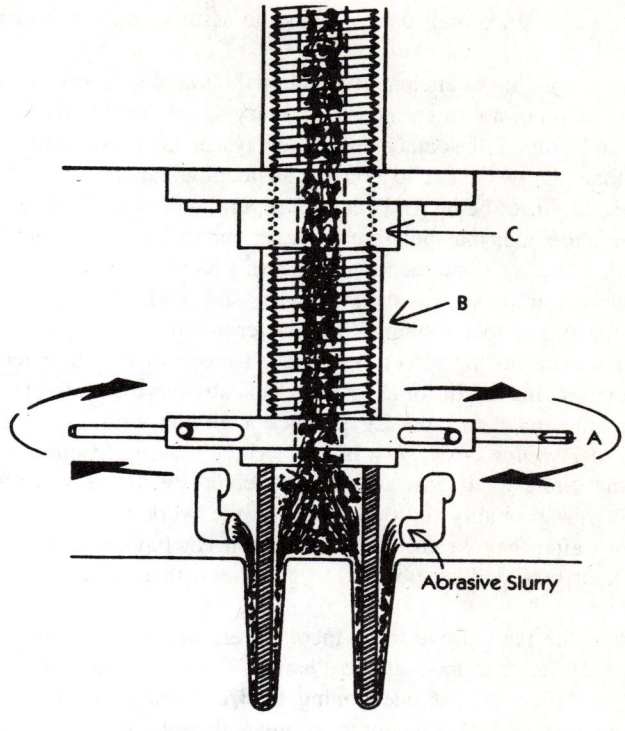

Ultrasonic Drilling of Egyptian Granite

FIGURE 8. A simple illustration showing the mechanical components, without the ultrasonic coupling, needed to drill the Egyptian granite.

implications of the data contained in this paper are quite substantial, and no doubt there will be controversy. One argument that could be leveled against the conclusions reached here may be the lack of certain kinds of evidence. Where are the machines, for example?

The tools that were used to build the more than 80 pyramids in Egypt have never been found. The few copper implements that have been uncovered, even if they were adequate to explain the evidence, do not begin to represent the number of tools of that particular type that would have been used had every stonemason who was supposed to work on the pyramids owned one or two. In the Great Pyramid alone there are an estimated 2,300,000 blocks of stone weighing between 2½ and 70 tons each. That is a mountain of evidence with no tools surviving to explain how it was built.

In the Great Pyramid there are mysteries yet to be explained that will strongly confirm my reasoning. Even though the tools and machines have not survived the thousands of years since their use, we have to assume, by the evidence, that they existed.

An understanding of ancient civilizations' level of technology shouldn't hinge on the preservation of a written record for every technique that they had developed. The "nuts and bolts" of society do not always make good copy. A stone mural will more than likely be cut to convey an ideological message, rather than the technique used to inscribe it. And yet after several thousand years, an interpretation of the artisan's methods may be more accurate than an interpretation of his language. The language of science and technology doesn't have the same freedom as speech.

But we are impressed with languages, and tend to believe if something is spelled out for us on paper. Champollion's interpretation of the Rosetta Stone seems to have had a more lasting effect than Petrie's study of the engineering aspects of the Egyptian monuments. In some writings, it is obvious that Petrie is totally ignored when claims are made that the Egyptians cut granite with copper chisels. When *Pyramids and Temples of Gizeh* was first written, it created quite a lot of interest in "dilettante circles." It was also well received by the press at that time. The *Saturday Review* probably didn't know that the work would still be an issue a hundred years after they wrote: "Mr. Petrie's survey having been made public . . . all future theorizers will be obliged to grapple with a series of incontrovertible facts."

Whether the facts prove to be incontrovertible or not, there's sufficient evidence to justify another look at the Pharaoh's stonemasons. In considering this interpretation, though, an unquestioning total acceptance could be as bad as an unquestioning rejection. It is meant to stimulate thought, discussion and, hopefully, a sincere objective study of the artifacts in question by anybody with the resources to do so.

George W. Harper

A LITTLE MORE POLLUTION, PLEASE!

OCTOBER 1986

AROUND 1970, it became fashionable to worry about various forms of pollution and the adverse effects they might have on future life. A few of us were doing that long before the fashion came, and some are still doing it after the fad waned. "Greenhouse effect" has become a household phrase. this article points out that even in righteous concern there is danger, for ecology is by no means a simple thing, and the effect something "obviously" has may be only a small and misleading part of the overall picture.

George W. Harper was a futurologist before the term came into common use, with a particular propensity for questioning the currently popular wisdom and a disturbing habit of being right with his unorthodox predictions. His education (with strong components of philosophy and history as well as astronomy and astrophysics) and a highly diverse occupational background enable him to look at the future from many points of view and see connections that might otherwise escape notice. He described a Mars heavily cratered by past vulcanism years before space probes showed the same picture, and he advanced a view of the outer Solar System that is only now becoming widely accepted. When he warns of the dangers of hasty solutions based on simple views of complex problems, the proponents of those solutions might do well to stop and listen. ■

IT is recorded that during the waning days of the last century, a modestly prominent middle-aged astronomer had to be pulled from atop a venerable geologist who obstinately persisted in arguing that geological evidence proved the Earth must be at least 300 million years old. Astronomy had already established that the sun could not possibly have flamed more than 75 million years, and for an obviously senile septuagenarian geologist to maintain a contrary opinion was an intolerable affront to the primacy of astronomers everywhere. To question an astronomer's professional opinion was heresy!

Modern astronomers are somewhat less violent but frequently no less complacent of their wisdom. Thus we find a recent diktat which ordains that if the human species does not promptly desist from spewing heat and particulate pollutants into the atmosphere the mean world temperature will rise about 9 degrees over the next century, melting the polar icecaps and flooding cities all over the world.

Just how we are to accomplish this in the face of colder winters and a rising human population is nowhere specified . . . a little genetic engineering so we all grow fur coats, perhaps . . . but the warning hardly qualifies as startling. In recent years a number of highly respected astronomers have pointed to Venus and gone on record warning us of the hazards of a runaway greenhouse effect, where the buildup of atmospheric CO_2 traps heat from the sun and makes the world uninhabitable. The 9-degree prediction is simply an extension and refinement of the greenhouse warnings, which focus primarily on the release of human-produced heat into the atmosphere. Both the long- and short-term perils are part of the pollution syndrome which has become so popular over the past few years.

Mention pollution to just about anyone and we are assured of the knee-jerk reaction: "Pollution! Ugh, that's awful. We've got to stop it before it kills everyone!" Pause to think a bit about this response and its pavlovian character starts to become disturbing. Anything capable of evoking so automatic a reaction is also susceptible of being misused . . . which recalls the remark by one of our more illiterate TV commentators who, a few years ago, scornfully pointed out that in leaving excess equipment behind on the Moon we were "now no longer content merely with polluting the Earth but have started polluting the entire universe with our litter!" It is amazing how minuscule the universe becomes when challenged by a news commentator. It is on a par with the remark by another commentator who opined that "Now that we've conquered space its time we turned our efforts to solving the problems here at home!"

Continuing in the same vein, anyone who wishes to liven up a dull party might try dropping a quiet comment that a little radiation is good for us. This is positively guaranteed to lead to a screeching confrontation by three or four individuals who are utterly convinced that radiation means death. "Radiation is pure pollution! Any amount is too much!" Try to explain that without nuclear radiation from the sun life becomes impossible on Earth, and they gasp, struggle briefly with an alien concept, and dismiss it as something wholly irrelevant. Observe that in burying spent nuclear waste we actually are returning less radioactive material to the ground than was originally taken out, and they stare blankly. They are so caught up in their fantasy of pollution they cannot grasp the elemental idea that when we extract energy from something, we necessarily wind up with less energy than we started with. We may increase short-term radiation, but no matter how we want to cut it the sum total of radiational waste is less than was initially extracted. Sorry about that, but it's the truth.

Sadly, none of this makes any real difference. The word "pollution," as society currently uses it, has become virtually meaningless. It is now primarily a pejorative to describe anything the speaker dislikes. If you don't like dogs, then call their leavings "pollution" and run them out of town. If you dislike humans, call our waste "pollution" and demand we stop breathing. Dislike rock music? You shout "noise pollution" and call for its elimination. Not long ago, authorities of a Pacific coast community actively considered fencing off several hundred square miles around a mountain reservoir and killing off the animal life to prevent "pollution!" That this would produce an ecological catastrophe was not even considered. "Pollution" has become a buzz-word so broad in content, it now has at best only marginal significance. It has been polluted by pollution-criers!

As we see it, there are so-called "pollutants" that it's highly desirable or essential to eliminate from the environment. There are other "pollutants" which are iffy and probably should be reduced in the environment, and, at least in the case of "air pollution" there is one which, if we successfully eradicate it . . . or even reduce it significantly . . . conceivably might lead to global catastrophe and possibly to the effective destruction of the human species!

We suspect even the chaps who today scream most shrilly about the need to eliminate air "pollution" would be a trifle disconcerted if they had to do their yelling from the bottom of a mile-thick ice sheet. It would be ironic justice, but the rest of us would have to suffer, too. And this is not so nice. It is entirely possible that a significant reduction in the air "pollution" index will lead directly and very quickly to a new glacial age which would smother our cities and farmlands under a massive sheet of ice.

Does this mean we question the possibility of a greenhouse effect? Are we flying in the face of observational evidence and denying its reality? Not in the least. Venus exists, so does greenhousing. But we take exception to the facile presumption that these facts are directly tranferable from Venus to Earth. Venus lacks water. Earth obviously does not. Venus is only 70 percent Earth's distance from the sun, which means it receives roughly twice as much heat per square cm. of surface as Earth. Earth, with its massive oceans, has evaporation, and a buildup of atmospheric moisture in the form of clouds will reflect solar heat rather than absorbing it, thus largely neutralizing the effects of any manmade heat or CO_2 excess that may be present. In short, while it is theoretically possible that an excessive buildup of carbon dioxide in the atmosphere could trap enough heat to melt the polar ice caps and flood coastal areas around the world, we doubt as a practical matter whether there is a realistic likelihood of creating a Venus effect.

But suppose we postulate the possibility of greenhousing on Earth, just to see where it leads us. We burn fuel to warm our homes and run our transportation systems and factories. The fuel burned (except for hydrogen, nuclear, and fuel cell systems) releases CO_2 into the atmosphere, which leads to a buildup of heat, which in turn leads to a general melting of ice caps and a dramatic increase in the percentage of Earth's surface water cover . . . melting the ice caps would add roughly 10 percent to the existing land-to-water ratio, all of it in the form of shallow, evaporatively efficient tidewaters. The same heat that leads to the melting of polar ice caps also tends to increase evaporation and creates dense cloud layers which reduce the intensity of the solar radiation by reflection.

With the generally warmer temperature we need no longer use so much energy heating our homes, so the rate of CO_2 release into the atmosphere slows. In the meanwhile, the shallow tidewaters become prime breeding grounds for vast quantities of algae which absorb CO_2 and release oxygen. Oceanic CO_2 absorption completes the picture, and conditions stabilize, probably well short of a complete ice melt. The overall effect is a partial melt of the ice caps and some flooding of coastal areas around the world. But there is no Venus effect, and while the consequences may be rather uncomfortable, the human race would survive. We might have ourselves a steamy, carboniferous-era climate where glimpses of the sun are rare and torrential rains sheet down daily, but the Venus effect per se does not occur.

Now suppose we look at some of the possible hazards of a significant reduction in air pollution. Astronomers may point to Venus, but geologists can point to Earth. Over the last several million years our planet has endured a whole succession of ice ages, when mile-thick glaciers perched atop what is now Chicago, when what is now Miami was 100 miles inland, the Gulf Stream flowed eastward south of Cuba, and icebergs ground against the Bahamas peninsula and clogged the California coast.

The last five ice ages appear to have come at approximately half-million year intervals, with the time between the end of one age and the onset of the next roughly 25,000 years. A look at the calendar therefore suggests we are about 10,000 years overdue for our next ice age! More pertinently, a study of Earth's climatic history over the past 7,000 years argues that we may actually have entered a new ice age around 6,500 years ago, but for some reason it was aborted.

The evidence is moderately convincing. Roughly 6,500 years ago, Earth's climate was halcyon. The snow line in Scandinavia was above the 8,000-foot level in many areas. Deciduous trees grew all the way to the arctic circle. The Sahara Desert region received adequate rainfall, and the peoples of the Badarian culture roamed throughout the Saharan grasslands. The evidence for all this is decisive. Badarian artifacts are found strewn over areas where no rain has fallen in centuries. Satellites photographing the African desert have revealed the beds of several former rivers, all of which save the Nile had dried up by about 6,000 years ago.

By 3,500 years ago, the tree line in Scandinavia had dropped to around 6,000 feet and deciduous trees were in full retreat from the circle. Climate throughout the world was changing. Temperatures had started cooling, and the snow lines on northern mountains crept lower every year. Growing seasons in Europe were shortening, and the winters even in Italy and Greece became a little more severe with every passing season. The fertile lands of Samarkand cooled and dried, and it became a desert.

A little less than 2,000 years ago the rains along the northern coast of Africa largely ceased, so today the erstwhile "breadbasket" of the Roman Empire is hard put to provide food for its own population and is mainly desert. At the same time this was happening in Europe and Africa, the winters in central Asia had deteriorated to a point where hordes of nomad herdsmen from Siberia, made desperate by lack of food, erupted onto neighboring lands with sullen ferocity and a single-minded determination to survive. The refugees driven from their homes by the northern invaders migrated into Europe, where they became known as Huns, Avars, Vandals, etc., and promptly toppled the Roman Empire. Since the victors against Rome were the *losers* of the battle for the Asian steppes, we can imagine the calamity that would have befallen Europe had the victors continued their pursuit!

Only a couple of centuries after the Huns and Vandals, the advancing cold made it no longer possible for Scandinavia to support its population. New waves of migration led to the Vikings and their conquests of Normandy, England, and Sicily and their discovery and settlement of Iceland and Greenland.

Thermal "pollution" from its volcanoes and geysers kept Iceland habitable, but the Greenland settlers had no such luck. The colony got off to a promising start, and several settlements were established. A bishopric was created and the population grew to three or four thousand. But nothing could stop the ice. Each year grew a little colder. Every year the growing season was a bit shorter. Every year the ice sheet encroached farther onto the pastures and fields. Massive icebergs began clogging the ocean and prevented ships from stopping by. Greenland vanished from the European horizon for over 150 years. When the next ship arrived, there was no one there to greet it. The last survivors had died over a century earlier, and in most places the snow and ice completely covered their old farmsteads.

Less than a thousand years ago, the cold was whistling down on the heartland of Europe. The Seine River froze solid in Paris, and wolves howled in the streets of the city as they fought the Parisians over the bodies of the dead. Ice formed regularly on the canals of Venice, and Rome experienced a succession of bitterly cold winters. Significantly, this same period saw the destruction of the Anasazi culture in the American Southwest. This has already been associated with a prolonged drought and a succession of abnormally frigid winters, so we can be reasonably certain we are dealing with a worldwide phenomenon rather than a localized aberration.

But this was to prove the high-ice mark. Around 1200 A.D. the weather began reversing itself, slowly at first then with increasing speed as Earth experienced a new warming trend which has lasted some seven centuries and only around 1950 started showing signs of turning cold once more.

There are several ways of accounting for this trend. Since knowledge of the fine detail of glaciation periods is scanty, we might call it a perfectly normal part of the glacial process: a preliminary cooling which lasts about 5,000 years, followed by a respite of perhaps 700 years before going into a second cooling era with other brief interludes to follow. This is undoubtedly a legitimate way of interpreting the evidence. If correct, we may assume Earth is in for a new cooling spell which will outdo the last, and humankind is in for some very nasty climate over the coming millennia.

A second way of accounting for the warming trend is more interesting. When the last glacial age started half a million years ago, man was something of a newcomer on the scene. The human (or proto-human) population was probably no more than a few million, with the majority still living in semitropical regions. When the glaciers moved south, those of our ancestors who were confronted by the advancing ice retreated before it. A few campfires were lit, but that was the limit of human response.

In contrast to the last ice age, by the year 1200 A.D. the human population of Earth probably stood at roughly 500 million, with the majority living in the northern hemisphere. Property and real estate rights were established institutions, something to be defended to the bitter end. A farmer does not desert his family plot without compelling reasons, and cities such as Rome, Athens, Paris, London, or even Moscow are not lightly abandoned. When cold comes you light a fire and tough it out. If it gets colder, you add another log and continue to sit tight. By 1200 A.D. the Earth's cooling trend was countered by the smoke from millions of fireplaces, all belching heat, carbon dioxide, and soot into the atmosphere. Cold "pollution" was met by heat, soot, and CO_2 "pollution." The cold lost and Earth gained a respite.

Which of the two alternative explanations is correct? Or are perhaps both correct, with a natural respite being further augmented by human-caused air pollution? We lack the data to permit an absolute conclusion—partly because no one has looked for it—but there are some fascinating elements to consider.

Earth's weather is narrowly balanced between extremes, and a change of three or four degrees in the mean world temperature is enough to start an ice age, or end one. This is known data. We also know that even small cities spew enough heat into the air to create micro-climates overhead. It is the same principle as that used by Florida citrus growers when a freeze looms. Smudge pots cannot cope with heavy frosts, but they are quite capable of muting the effects of a modest one. Of special significance is the fact that we do not get the same effect if we merely deploy a batch of electric heaters. "Pollution," or smog, emitted by the smudge pots is essential if the remedy is to work. It creates an artificial cloud of sooty smoke which reflects the heat back and contains it over the citrus grove. Without the smudge the heat simply radiates off into space and is quickly dissipated. With the smudge, the same units of heat energy are multiply reflected and confined.

By clear implication, the medieval cities, small though they were by comparison with modern metropolitan areas, were still significant in creating islands of resistance to the cold and thus in modifying the prevailing weather patterns. As the population increased and more smokestacks belched out their "pollution," the climate was further eased. This effect *must* occur regardless of any natural slaking of the cold. In short, even if the recession of the mini ice age was a natural consequence of the pattern of ice ages, we can still be certain that it was further ameliorated by human action. The only question is whether human activity is sufficient in itself to account for the reversal of the cold.

Here again is uncertainty. We lack the data to come up with an unambiguous answer. But the last half-century has provided some disturbing evidence. The British Isles are a good place to begin. For centuries London has been famed for its impenetrable fogs which swath the city for days on end during winter months. Starting around 1950, the English began switching from sooty coal pots to electricity for heating. Much of this power comes from clean nuclear sources, the rest from emission-efficient central generating systems. By 1960 the English countryside had become mostly soot-free and the notorious London fogs were largely a memory. But so were the warm, rainy winters which used to characterize the British climate. Instead, the air is clear, the ground heat quickly dissipates into space, and temperatures have become abnormally cold, with heavy winter snows commonplace.

France too is feeling the effects of a newly "pollution-reduced" environment. Clean, centralized energy has lowered air pollution and dropped temperatures. In the winter of 1984–85 a blizzard swept down on the Riviera and a number of campers froze to death on the Mediterranean coast—which is roughly equivalent to a snowstorm blanketing Honolulu! It is not something that just happens.

Paralleling the deepening cold in Europe is drought in Africa and elsewhere around the world. Lack of rain has led to a systematic southward march of the Sahara, commencing around the late 1950s and continuing uninterruptedly every year since. Then there is the succession of disappointing harvests which have afflicted the Soviet Union. Our propagandists cheerfully attribute the failures to inherent defects in the Soviet system, but while this may be a contributing factor the primary cause has been a steady worsening of the weather and frequent droughts over the past three decades.

This too may be connected with a reduction in suspended air particulates and atmospheric CO_2. Water droplets form most readily when aquaphilic "pollutants" such as soot particles are adrift in the atmosphere to serve as condensation nuclei . . . witness the occasional successes of rainmakers seeding clouds with silver-iodide crystals. It is therefore entirely possible that in our obsessive urge to eliminate air "pollution," we may have been the direct cause of the drought-induced famines now killing tens of millions of people throughout the world!

Nor has the United States been immune. As we have cleaned up our air, the climate has worsened. We may have made southern California marginally more endurable, but on the other side of the coin we find massive winter freezes devastating the fruit crops of Florida, whole successions of hurricanes lashing the eastern and southern seaboards, and droughts, abnormal freezes, heat waves, and floods pestering the Midwest. Every year for the past decade has seen new records for cold being set in the southlands. When Atlanta gets nearly a week of zero weather—as it did during the winter of 1984–85—we have to call it unusual. Winter snows all over the nation have been getting heavier every year for the past decade, and the proportionality between precipitation in the form of rain and precipitation in the form of snow has been edging in the direction of snow. Keep at it a little longer, and the whole nation may find itself in deep trouble.

We concede the possibility that all these phenomena would have occurred no matter what we did. Perhaps the mini ice age briefly hesitated for reasons wholly unrelated to human activity and is now getting down to serious business. For that matter, possibly the climatic changes over the past 50 years are no more than minor statistical glitches which will straighten out over the next century. At the same time, we should not ignore the fact that a number of profound changes in Earth's climate have appeared only *after* we set about deliberately cutting back on air pollution.

Some of these changes, for example, the disappearance of London fogs, are demonstrably a consequence of the reduction. Others, such as the sub-Saharan drought, are less certain. But the indisputable fact remains that clear, unpolluted skies lead to rapid thermal loss, while clouded skies make an excellent thermal blanket and retard heat dissipation. Clear, unpolluted skies are inefficient at transporting water over land. Optimum efficiency calls for considerable "pollutants" to serve as condensation nuclei and to help avoid the extremes where we either get no rain at all or suffer typical desert downpours where the clouds dump their loads all at once to create enormously erosive flash floods below.

We may even accept an argument that our relatively small reduction in air pollution is insufficient to cause climatic changes of the magnitude experienced in recent years. It is still a scientific certainly that at some point along the line the effort to eliminate air "pollution" *must* lead to colder winters and less rain. It is the other side of the coin to the Venus greenhouse effect and is inherent in the nature of the statement.

The process is three-pronged. For one, we have a clearer atmosphere where solar heat reaching the Earth is reradiated back into space every night. This means less heat is available over the yearly cycle and the average annual ground temperature must be reduced. It implies a smaller underground heat buildup, so when the first snows of winter fall they quickly deplete the available ground heat and the snow remains on the ground longer.

The second prong develops when the reduced aggregate usable solar heat leads to persistently colder arctic and antarctic temperatures and the expansion of the snow blanket. Snow is an extremely efficient reradiator of thermal energy. Increase the snow cover even temporarily, and we find even more of the sun's heat being reflected back into space. This means less heat is available to warm Earth.

The third develops when coriolis effects slide evaporated moisture from the tropic oceans toward the north and south polar regions, where at least a portion of the moisture is deposited as snow. The moisture which descends onto the arctic as snow is not going to be deposited elsewhere as rain.

Combining the three factors we arrive at a pattern where world temperature reduction tends to exponentiate. Most of the atmospheric moisture falls as snow, so the non-glaciated areas become arid. Confinement of water in the form of snow and ice leads to a lowering of the sea level. This latter reduces the water surface area available for new evaporation, which thereby increases the drought in the unglaciated areas. The system becomes self-perpetuating over a long-term cycle.

Now let us suppose, purely for the sake of argument, that for the past 6,500 years our world has been making an all-out effort to whip up a new ice age, but for the past few centuries man has unwittingly forestalled the cold by his use of fossil and wood fuels. We have not only held the cold at bay but have actually pushed it back a little. But it is a fragile victory where even a fractional percentage point of change in the index of air "pollutants" can tip the balance. If our hypothesis is even partially correct it may easily turn out that the one or two percent reduction in worldwide air "pollutants" achieved over the past 35 years will lead to a runaway refrigerator effect rather than the greenhouse effect so often cited by astronomers.

Should this happen one thing is certain: neither Russia nor the United States will sit back supinely and perish as the ice descends. Great nations do not die quietly, and when it comes to sheer survival, no national leader will flinch from employing the most extreme tactics . . . which today means nukes. Thus our deep and abiding suspicion that in our eagerness to abolish the demon "air pollution," we may be guaranteeing a far worse catastrophe not far down the line.

But let's be positive for a minute. Perhaps the elimination of air pollution from the environment will not cause a new ice age. Possibly all this is unjustified pessimism on our part. We still have ample cause for concern, a concern arising from the fact that no one has bothered investigating the possibility. It is evident that the prophets of doom have evolved their mythology on the premise that Earth is static and today's climate is all there is. Ice ages are gone and forgotten, so there is no point in worrying about bringing on a new one. In the classical sense of the word they have advanced a half-baked idea based on the false-to-fact premise that Earth is static, and have failed to study either the implications or the long-term ramifications of their idea. And this is frightening! In today's technological society one or two men of scientific prestige may make a hasty or ill-conceived statement whose consequences a century or so down the line can be utterly catastrophic.

It has been said with some justification that the most dangerous of all ideas is the noble one, the sort of idea aimed at engaging the passions rather than the mind. "Pollution" is merely the latest of a whole series of such ideas. Humankind desperately needs a system to get at least a few people thinking about the consequences of these noble actions and ideas. There ought to be some organized group whose special function is to say, "So get rid of pollution. It's a great idea; now what nasty side-effects can throw a monkey wrench into the works?" This business of looking for possible damaging outcomes to glorious experiments is almost entirely lacking. It has been left to professional nay-sayers and panic-mongers who usually work from wholly inadequate data to arrive at absurd conclusions.

But the analysis of consequences is too important to be left in the hands of penny-dreadful national newspapers which base their sales on tales of women giving birth to alligators, of visits by the ghosts of dead movie stars, and the predictions of psychics who tell us of Mu, Lemuria, and Atlantis. We also feel it is entirely too important to be left to the mercies of politicians who look only as far as the next election or of business managers concerned solely with next year's profit and loss balance.

It would be nice to believe our university system could take up the slack and fill the niche, but since WWII most have become so beholden to political and corporate funding they do little more than rubber-stamp the decisions of their masters. What is really needed is a scientific *Consumer's Digest*-type operation. Sadly, none is anywhere on the horizon.

Part Two

THE UNIVERSE

WE LIVE IN

Richard Matzner,
Tsvi Piran, and
Tony Rothman

DEMYTHOLOGIZING

THE BLACK HOLE

SEPTEMBER 1980

FEW objects in astronomy have captured the popular imagination as strongly as the black hole, a most peculiar phenomenon predicted by the general theory of relativity and not yet unequivocally detected in the real universe. Strange as they are, black holes are not *quite* as bizarre as the folklore that has grown up about them. This article, longer than *Analog*'s readers usually prefer but resoundingly voted their favorite of 1980, attempts to set the record straight.

The coauthors are uniquely qualified to do so. When they wrote the article, all three were actively engaged in research in this field. Richard Matzner was an associate professor in the Center for Relativity at the University of Texas; he is now its director. Tony Rothman was then Matzner's graduate student, doing a dissertation on primordial black holes. He received his Ph.D. in physics in 1981 and has since held postdoctoral fellowships at Oxford University, the Shternberg Astronomical Institute in Moscow, and the University of Cape Town, South Africa. He is also active as a writer and a musician. Tsvi Piran did a dissertation on astrophysical processes near black holes for a 1976 Ph.D. from the Hebrew University in Jerusalem. He has since done research at Oxford, the Center for Relativity, and the Institute for Advanced Study in Princeton. ■

INTRODUCTION: "TRUTH IS FOR THE MINORITY"

With the release of the Disney catastrophe, general interest in black holes has peaked. The release of this film also signals a critical overdose of misinformation to which the public has been exposed. We read statements like: "The pull of a black hole's gravity is so strong . . . Time is stopped and space does not exist. . . . [A black hole's discovery] would unravel the mystery of both the universe's creation and eventual destruction."

Such blatant idiocy induces the public conception of black holes as monsters which gobble up all the matter in the universe, as miracle workers which can solve all our energy problems, as gateways to other universes, and as time machines. This conception is profoundly misplaced. The same theory which predicts the black hole's existence also predicts that each of the preceding properties has severe limitations or does not occur at all. The very existence of black holes is itself debatable; within our own galaxy, only one not-yet-conclusive candidate for a black hole has been found to this date, the X-ray source, Cygnus X-1.

Thus, it strikes us as bordering on the ridiculous to use black holes as an explanation for every property of known space. Of course, there are mistakes and there are mistakes. Some involve subtle points, and physicists advance their own field only by making lots of them. The layman cannot be faulted for doing the same. Nonetheless, most of the nonsense written about black holes stems from an ignorant exploitation of a sublime idea and a lack of interest in the pursuit of knowledge. As we will see, the theoretical properties of black holes are in themselves so remarkable that there is no need to exaggerate them in an attempt to capture the public's attention. Bearing this in mind, we now examine some properties of black holes—without exaggeration.

CLASSICAL BLACK HOLES AND CANONICAL MISCONCEPTIONS

Visualize a black hole. Most of us, encumbered by the limits of imagination, will visualize a small, black sphere floating in space among the stars. We probably think of this ball as a highly compressed solid, something like cold iron but unimaginably more dense. Unimaginably high density, we assure ourselves, produces an unimaginably great gravitational field. We further imagine the field to be so strong that all surrounding matter is pulled into this tiny sphere, never to escape again. Light itself cannot avoid the same fate; fleeting, ephemeral, yet once light enters this strange object it is trapped forever by gravity. Thus, the "black hole": absolutely black since light cannot be reflected from it to show its existence.

The question is, is this picture a description of anything? The answer is not straightforward but requires more precise concepts, caveats, and "yes buts." In attempting an answer, one should first keep in mind that relativists, peddlers of gravitational theories, distinguish between several types of black holes. There is the basic, Schwarzschild black hole, which is spherical, electrically uncharged, and does not rotate; there is the Kerr black hole, which rotates and is not spherical; and there is the Reissner-Nordstrom black hole, which is spherical and non-rotating but contains an electric charge. (The holes are named in honor of the mathematicians who worked out their theoretical existence.)

These three types, without additional complications, are lumped under the heading "classical black holes" to distinguish them from "quantum black holes." A quantum black hole is any black hole, including one of the above types, which it is necessary to take into account the fact that light, for instance, consists of indivisible units called quanta. For light the quanta are photons: for the gravitational field itself the quanta are gravitons. Thus we can have quantum Kerr black holes, quantum Reissner-Nordstrom black holes, and quantum Schwarzschild black holes. But for now we limit ourselves to classical black holes.

Evidently, the above mental picture corresponds—more or less—to the basic Schwarzschild black hole. However, the emphasis in the previous sentence is on the "more or less," specifically on the "less." We will now begin to give a more accurate description of a classical black hole, keeping in mind that specific details may vary from one category of hole to the next.

The classical, astronomical picture of a black hole is one of a remnant left over by the collapse of a massive star; the examples typically used have about ten solar masses. The escape velocity from the surface of the black hole exceeds that of light; indeed, this is the definition of both a "black hole" and its "surface." The surface of a black hole is called the *event horizon*. Now, we know that no physical object can move faster than light, so nothing whatsoever, having fallen across the event horizon of a black hole, can come back out through that horizon.

A ten-solar-mass black hole has a radius of about 30 kilometers, roughly the size of New York City. It is this typical example of a small, collapsed object with a gravitational field so strong that not even light can escape, which has conjured up the vision of black holes as extremely dense objects which grab anything in the vicinity. In fact, the density of the ten-solar-mass hole (density is the mass of the hole divided by the volume enclosed within the event horizon) is of order 10^{15} grams per cubic centimeter. This seems a very high density by everyday standards (the density of iron is only about 8 grams per cubic centimeter), until we realize it is comparable to the density in the nuclei of atoms. Each one of us is composed of particles of this sort of density.

In any case, a black hole does not have to be so dense. The basic black hole equations show a very simple relationship between the size and mass of a black hole and the density. As the radius of the hole or its mass is increased, the density goes down. Thus, by making a black hole large enough or massive enough, we can make the density as low as we want. Actually, there is no reason we could not make a black hole out of air. Such a hole would have a radius of about 30 billion kilometers, roughly ten times the size of our solar system. If one entered this black hole, one would hardly feel a thing, but after a few days life would become uncomfortable—as one approached the singularity.

While we will not talk much about singularities in this article, we should mention that the singularity in the center of the black hole is the place where all the matter eventually ends up. The density at this point is infinite, which introduces a "yes but" into the above remarks. The density we have been discussing is the *average* density of the hole, and, strictly speaking, one can only talk about the average density from outside the horizon.

At the surface of the air-bag black hole, the gravitational acceleration would be about 100 times the acceleration we feel on the surface of the earth, or roughly the same as the gravitational acceleration on the surface of the sun. A larger black hole, made out of hydrogen, would have an even lower surface acceleration. We see, then, the gravitational acceleration of a black hole is not always overwhelmingly large.

If the gravitational field is so weak, the question immediately arises, why can't one escape by firing a rocket engine. The answer is somewhat tricky. We know that by accelerating even at very low accelerations, say .1 gee, we can eventually reach huge velocities. Similarly, even the weak acceleration produced by our air bag will eventually accelerate objects to high velocity. In fact, by the time an object has fallen to the event horizon of any black hole, it is moving at the speed of light, inward. If the falling object wants to remain even stationary at the horizon, it must then move with the speed of light, outward. The principle of relativity says that nothing can move faster than the speed of light. Therefore, there is no escape. Acceleration is somewhat irrelevant to the problem; the speed of light simply cannot be exceeded.

Fancy Free: Orbits Around Black Holes

Related to the idea that a black hole possesses a strong gravitational field is the misconception that nothing can get remotely near the hole without being gobbled up. A good illustration of this nonsense is in the Disney film where the ship, the *Cygnus*, seems to require an antigravity field to prevent it from falling into the black hole. The filmmakers ignore the fact that, at distances greater than about 10 times the radius of the black hole, ordinary orbital mechanics—known since the time of Newton—is applicable. For example, if the sun were suddenly replaced by a black hole of equal mass, the orbits of the planets would not change by the width of an ant's eyebrow. Admittedly, it would get dark, but that is another story.

This brings us to the first important rule of black hole orbital mechanics: *At large distances, the fact that we are in orbit around a black hole is irrelevant.* We may consider the black hole to be a spherical mass concentration producing an ordinary, Newtonian gravitational field, like that of the Earth or the sun.

As the orbital radius approaches 10 black hole radii (300 kilometers for the ten-solar-mass case), general relativistic effects become very important. The proverbial *curvature* of space and slowing of time come into play. Such distortions of space and time manifest themselves in such effects as perihelion shifts in noncircular orbits. To understand perihelion shifts, we recall that Newtonian orbits are steady ellipses around the central body. The satellite's point of closest approach, the perihelion, remains at a fixed point in space. We say, in this instance, space time is flat or Newtonian. (Figure 1a.) When curvature of spacetime is more significant, the point of closest approach pivots around the central body with each orbit of the satellite. (Figure 1b.) This pivoting is called a "peri*helion* shift" when speaking of orbits around the sun, a "peri*astron* shift" when speaking of orbits around stars in general, and a "peri*barythron* shift" when speaking of orbits around black holes. ("Barythron" is the Greek name for a deep pit in Athens into which condemned criminals were thrown.)

Because the shift is a cumulative, continuous effect, it can be detected even in satellites far from the central body, if a sufficiently long time is spent on the observation. For instance, Mercury's perihelion shift is about 42 seconds of arc per century, a very small effect indeed. We can say that, as far as Mercury is concerned, the spacetime curvature caused by the sun is hardly noticeable. Spacetime is very nearly flat. Close to a black hole, on the other hand, the peribarythron shift becomes very important. At ten black-hole radii, it amounts to about 70 degrees per orbit!

Even closer to the black hole, circular orbits become unstable. A small deviation inward leads to a continuing spiral into the hole. For a Schwarzschild black hole, the point of instability comes at 3 black-hole radii (i.e., at 2 radii from the surface). This does *not* mean anything which falls within 3 radii of the black hole is irretrievably sucked in. One may still swoop down from a very large distance, down to 2 radii, and return to infinity, just as a comet approaches and then recedes from the sun. And this approach can be made without engines, again, like a comet. If rockets *are* employed, one can come almost all the way down to the Schwarzschild radius, i.e., the horizon, and out again. Alternatively, one can continue to orbit around the hole below 3 radii; but, in this instance, rockets must be fired to maintain position. What is *not* allowed in this region are free, uncorrected orbits like those of satellites and skylabs around the Earth.

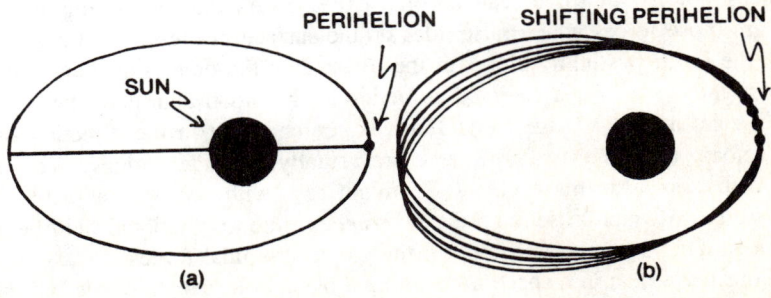

FIGURE 1. (a) A nonprecessing elliptical orbit. The perihelion, or closest approach to the sun, remains at the same point in space after each orbit. (b) A precessing orbit. The perihelion changes position slightly after each orbit around the sun.

In the Disney film, the featured hole was not a Schwarzschild, but a rotating or Kerr hole (even though the computer graphics shown during the credits were mistakenly those for Schwarzchild). For a Kerr hole, the point of the last stable orbit depends on how fast the hole is spinning, but the results are comparable to the Schwarzschild case; instabilities set in between 1 and 9 radii. Thus, the *Cygnus* should not need "antigravity devices" until very close to the hole indeed. On the other hand, as far as 1,000 radii from the black hole, the *Cygnus* would be orbiting with a period of about one second. Admittedly, this may be why the antigravity device was posited in the first place—to dispense with orbits altogether. On the third hand, we doubt the filmmakers thought this far.

Since we have been speaking of orbits, it is appropriate at this time to introduce the second important rule of black hole orbital mechanics: *The principle of equivalence still applies*. This fact seems to have escaped the attention of almost all moviemakers and writers. The principle of equivalence states that any body in a free orbit or in free fall does not experience the force of gravity. We might say, "Falling free or orbiting 'round, equivalence says gravity not found." Examples of this are encountered in everyday life: When we dive off a diving board we feel weightless. When an airplane drops suddenly, those in it feels momentarily weightless. Astronauts in orbit around the Earth are *not* weightless because gravity has been turned off above the atmosphere; rather, they are falling around the Earth, continually diving off the board, if you will. Under these circumstances, the principle of equivalence says that gravity is not felt.

This is a very important point which applies to *any* situation near a black hole when rockets are *not* being fired: orbiting on a stable orbit; orbiting on an unstable orbit (when not correcting for instabilities); spiraling in; swooping down like a comet; or just falling in. In these cases, one does *not* suddenly feel heavy near the hole. On the contrary, one feels weightless, as if he were orbiting the Earth or diving into a swimming pool.

There is a complication to be introduced here. When an object comes close to a black hole, *tidal* forces can become extreme. As their name implies, tidal forces are those forces which raise tides on the surface of the Earth. Because one side of the Earth is slightly closer to the moon than the other side, the near side feels a slightly greater gravitational attraction to the moon than does the far side. Thus, we get a "tidal bulge"; the Earth is stretched out in the direction of the moon. (Some readers may know there are actually *two* tidal bulges. We do not pause to discuss why this occurs.) We might say, with fair accuracy, that tidal forces are those which arise from the *difference* in the gravitational field between two points. The greater the difference, the greater the tidal forces.

Consider a man in a spacesuit orbiting a black hole. He is in free fall, so by the previous discussion, he feels perfectly weightless. However, the feet of the astronaut are slightly closer to the black hole than is his head. Therefore, he experiences tidal forces: his feet are being pulled toward the hole more strongly than his head. As a result, the astronaut is stretched. One might think, because a man is so small, that the difference in the gravitational force between his head and his feet cannot be very large. After all, gravity does not decrease *so* fast over a couple of meters. This is not true. Near a white dwarf, neutron star, or black hole, tidal forces can be immense. If he is orbiting a 1-solar-mass body at a height of 10 kilometers, the tidal forces on our astronaut are approximately ten million times the force the Earth is, at this moment, exerting on us. That is, while the Earth is pulling us to its surface with a force which, by definition, is equal to our weight, the astronaut is being ripped apart by forces about ten million times stronger. This particular example has roughly the conditions presented in Larry Niven's story, "Neutron Star." It is, alas, ludicrous to think the hero could save himself by curling up into a ball at the ship's center. More likely, he would end up spread over the walls, the consistency of pink applesauce. Perhaps, we have estimated, if he initially started out as a piano wire for triple high C, he might have survived.

Still, a caveat is in order here. Near black holes which are large enough, like our air bag, tidal forces become totally insignificant, much less than even those tidal forces we feel on Earth. Thus, no shredding at all will take place near these holes until one falls close to the singularity. At the singularity, in all black holes, the tidal forces are infinite.

To sum up this section, we reiterate that it is the tidal forces which wreck spaceships near black holes, not the simple fact of strong gravity. And, as just mentioned, for very large holes, over about 10^5 solar masses, even this does not happen. As an astronaut orbits a black hole, he feels as weightless as if he were floating amid the clouds on a fine spring day. Near a typical black hole, though, his head is being wrenched from his feet by forces which make the bed of Procrustes amateurish by comparison.

Being and Nothingness: Black Holes as the End of Space and Time

Two astronauts are orbiting a ten-solar-mass black hole. Richard, having seen one too many bad science fiction films, decides to end it all by taking the fateful plunge. He jumps. Tony, curious to see the demise of his dissertation advisor, decides to clock Richard's fall to the event horizon. "Time is on my side," Tony chuckles to himself, but he has a surprise waiting for him. As Richard approaches the event horizon, he seems to fall more and more slowly. Tony knows this because Richard is carrying a green, flashing beacon. The time interval Tony measures between each flash of the beacon is becoming longer and longer. In addition, he is startled to find that the flashes are growing much redder and dimmer "as time goes by." Tony grows impatient, but to no avail. The fall seems to take forever. Tony dies of old age muttering, "*Veritam dies aperit*," but Richard has still not reached the event horizon. Tsvi arrives in his space shuttle to take over the observations but suffers Tony's fate. He too grows old watching Richard's beacon flash ever more slowly and redly. With his dying breath, he entreats, "Stand still you ever-moving spheres of heaven / That time may cease and midnight never come." Tsvi's descendants have no better luck. Richard fades away completely just as he reaches the horizon, after a truly infinite amount of time. The clock has stopped.

Richard, on the other hand, realizes, "Time and tide wait for no man." He does not notice his beacon flashing any more slowly than normal, nor does he notice it growing redder and dimmer. He reaches the event horizon after a perfectly finite number of flashes. From that point, he crosses the event horizon, although he does not realize he has done so, and continues his plunge to the singularity at the center of the black hole. Of course, Richard is ripped apart by tidal forces long before he gets there, but his dispersed atoms reach the dreaded singularity in a rather short amount of time—about 10^{-4} seconds as measured by his flashing beacon.

As well as a mild discrepancy between two clocks, there is a moral to this fable: Relativity is called relativity because relativity is truly relative. The question, "Does time stop at a black hole?" is meaningless as it stands. We can say, "To an observer in a spaceship, an object falling into a black hole takes an infinite amount of time to reach the event horizon." But we can also say, without contradiction, "To an observer falling into a black hole, the time required to reach the event horizon is quite finite." When posing relativistic questions, one must be careful to specify about whom one is talking, or else one runs the risk of lapsing into gibberish.

The slowing down of Richard's beacon-clock (as measured by Tony and Tsvi on the ship) and the reddening of the light are two aspects of the same effect. The curvature of spacetime associated with the gravitational field around the black hole actually causes time to flow at different rates. Just as the flashing of the beacon can be thought of as a clock, so can the oscillations in a light beam, or the movements of atoms in the beacon motor. *Everything* is slowed down from the point of view of Tony or Tsvi on the ship. The slower oscillations of the light are interpreted by Tony's eye as a reddening of the light, and since light is being emitted from the beacon at longer intervals, fewer photons (light particles) reach the eye per unit time. The combination of these effects causes the excessive dimming of the beacon.

Richard, however, falling into the hole, is subject to the principle of equivalence. (Falling free or orbiting 'round, equivalence says gravity not found.) He does not feel any gravity on him or on his beacon. As far as he is concerned, there is no gravity to slow down his flashes and everything proceeds as normal, with the exception of tidal effects.

It is important to keep in mind that all these effects occur around any gravitating body, the sun for instance. The only difference is in the magnitude of the effects, which will be much greater around a typical black hole than near the sun or the Earth.

To conclude this brief discussion of space and time near a black hole, we would have wished to comment at length on the quotation found at the beginning of this article, to the effect that, a black hole is a place where "space does not exist." This, unfortunately, has proven to be impossible because we have entirely failed to discover in that statement any meaning whatsoever.

The Cosmic Whirlpool: Kerr Holes and Penrose Processes

Present energy dilemmas have made popular the idea of extracting large amounts of energy from black holes. The attraction of this idea is not hard to see. We are all familiar with the large flywheels used by electric companies in their power plants. These huge flywheels store *rotational energy*. By coupling the flywheel to a generator, we are transforming the rotational energy into electricity for use in home and industry. In doing so, we have extracted the rotational energy from the flywheel and, as a consequence, it slows down.

Now, we have mentioned that Kerr black holes rotate, much like the above flywheels. The rotational energy of a rapidly rotating solar mass Kerr hole is about 10^{54} ergs. At the Earth's present rate of energy consumption, 10^{54} ergs would last approximately 10^{27} years, or about 10^{17} times the present age of the universe. This is a long time.

The question naturally arises, can the rotational energy of a Kerr hole be extracted? If it could, we would expect the black hole to slow down like the flywheel. When no further energy could be extracted, the black hole would no longer be a spinning Kerr hole; it would be a non-rotating Schwarzschild hole. In 1969, the British relativist, Roger Penrose, showed that extraction of the rotational energy of a Kerr hole is possible. Immediately after his suggestion appeared, others further proposed that the Penrose process might be used by an advanced civilization to tap the energy of black holes. From there, science fiction took over. The basic idea was used in *Gateway* by Fred Pohl. Indeed, one of us (T.R.) succumbed to the temptation to use the idea in his novel, *The World Is Round*. Unbeknownst to T.R., T.P. and others were at the same time proving how difficult the Penrose process was to implement.

To understand the Penrose process further, we must first talk in more detail about Kerr holes. The rotation of a Kerr hole causes a "whirlpool in space." This whirlpool is actually quite similar to an ordinary ocean whirlpool except that, instead of water whirling around, it is spacetime itself swirling around the black hole. If a space traveller is caught in this whirlpool, he is dragged around the black hole exactly as he would be dragged around the eye of the vortex if caught in an ocean whirlpool. If the space traveller wanted to remain stationary, he would have to fire his rocket engines to overcome the spacetime dragging. Again, this has a marine analogy. A swimmer must swim against the current in the vortex if he wishes to remain in the same place.

We should note that this dragging is not unique to black holes but, according to relativity, occurs around any rotating body. In fact, a team of experimentalists at Stanford, led by Francis Everitt, is planning to measure the dragging force caused by the *Earth's* rotation. This measurement will be carried out by a satellite to be launched by the space shuttle. The dragging caused by a tiny body like the Earth is really very small. While the Stanford satellite orbits the Earth, the gyroscopes on board will be tilted a slight amount by the drag. After a full year, the cumulative angle of tilt will be less than a second of arc, about the angle subtended by a penny as seen from a distance of a kilometer.

Although the effect due to the Earth is small, around a black hole the dragging can become enormous. In fact, beneath a certain distance from the hole, which is termed the "stationary limit," no matter how hard one fires his rocket engines against the current, the dragging cannot be overcome and one is inevitably swept around the hole. This notion can be made more precise. Consider an observer on a "space buoy" being dragged passively around the hole. To him, someone in a rocket trying to overcome the dragging will appear to be moving in the opposite direction. At the stationary limit, this rocket will appear to the observer on the buoy to be moving at the speed of light. From a space station far above, however, the rocket is just managing to fight the current and remain stationary, hence the name "stationary limit."

We recall the famous words of the Red Queen: ". . . it takes all the running you can do to keep in the same place. If you want to get anywhere else, you must run at least twice as fast as that." Unfortunately, one cannot run any faster than the speed of light. If she is unlucky enough to fall beneath the stationary limit, even the Red Queen will never be able to stay put and will be dragged around the hole along with space buoys, rockets, and everyone else.

The region between the stationary limit and the event horizon is called the "ergosphere." "Ergosphere" was coined by Wheeler and Ruffini from the Greek word "ergo" meaning "work." It is in the ergosphere that the Penrose process takes place. (See Figure 3 for the relationship between the horizon, ergosphere, and stationary limit.)

Consider a rocket orbiting in the ergosphere. It ejects a load of garbage *against* the current (like the Red Queen). Although this garbage is swept around the hole —since it *is* beneath the stationary limit—it is "struggling against the current." One can imagine that an object moving on such a "counterrotating" orbit would exert a braking force on the hole and therefore slow it down in the same manner as we slow a flywheel. Thus, some of the rotational energy is lost and is, in fact, transferred to the ship as a recoil effect. (Think of a gun shooting a bullet. The recoil is greater than normal due to the presence of the rotating black hole.) This energy would be manifested as a greater kinetic energy of the ship, that is, a higher velocity. The ship could then leave the ergosphere with more energy than it had to start with, to be used elsewhere.

However, the matter is not so simple. The ejected garbage will be captured by the black hole, adding its own mass to the original mass of the hole. Since $E = mc^2$, by losing the garbage we are losing energy to the hole. If the amount of energy lost to the hole is greater than the amount of energy gained by braking the hole, we have a net loss of energy. No extraction has taken place.

Nonetheless, if certain conditions are met, the energy balance will be favor-

FIGURE 2. An observer on a space buoy, being passively dragged around the black hole whirlpool, sees the Red Queen at the stationary limit running at the speed of light. From a space station far above, however, she is seen as just managing to remain at the same place. (Note: This drawing is not to be taken too literally. One does not actually see a spacetime whirlpool around a black hole.)

able. That is, if the garbage is ejected at sufficiently high velocity onto a counter-rotating orbit within the ergosphere, the net result will be an energy gain. Any orbit which meets these requirements is termed an "energy extraction orbit." We emphasize that they only exist within the ergosphere. Note also that it is the orbit which is important, not what we eject. Therefore, it makes sense to use garbage, since this eliminates waste disposal problems as well.

The Penrose process is best illustrated by the machine in Figure 4, adapted from the text *Gravitation*, by Misner, Thorne, and Wheeler. An advanced civilization builds a huge shell around a Kerr black hole. Space shuttles loaded with garbage enter the ergosphere. They eject their payloads in the manner already described, receive a giant dose of energy which boosts them to huge velocities, and return to the surface. They are caught in the arms of a giant generator which converts this kinetic energy into electricity for use over the shell.

Even if our supply of garbage is limited to one Earth mass, this is enough to

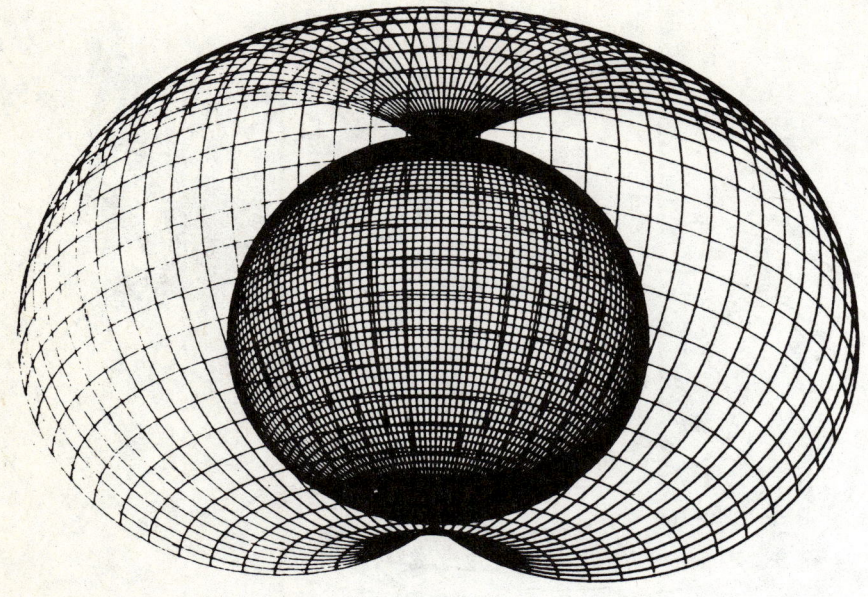

FIGURE 3. A computer-generated picture showing a cutaway view of a Kerr hole rotating at the speed of light. The outer surface is the stationary limit; the inner surface is the event horizon itself; and the region in between is the ergosphere. The view is from 26.5° above the equator, which accounts for the slight distortion of the inner sphere. We wish to thank Nigel Sharp for generating this picture. (Important note: All previous textbook views of Kerr holes in Boyar-Lindquist coordinates without the cusps shown here are incorrect.)

power the Penrose process for about 10^{21} years, or 10^{11} ages of the universe, not an insignificant amount of time.

Two technical details make this extraordinary picture somewhat less optimistic. The first difficulty is jettisoning the garbage onto an energy extraction orbit. The second difficulty is getting the boosted shuttle out of the ergosphere without being captured by the black hole. We can better understand these problems if we pretend we are on a shuttle, the *Penrosia*, whose mission is to go into the ergosphere, dump garbage, and return to the shell with as much energy as possible. The crew is fresh out of Starfleet Academy and so learns by the dangerous method of trial and error.

We have entered the ergosphere. Because fuel supplies are limited, the Captain has turned our engines off. The *Penrosia* is now being passively dragged around the black hole's whirlpool like a space buoy. Since we are in orbit, we feel weight-

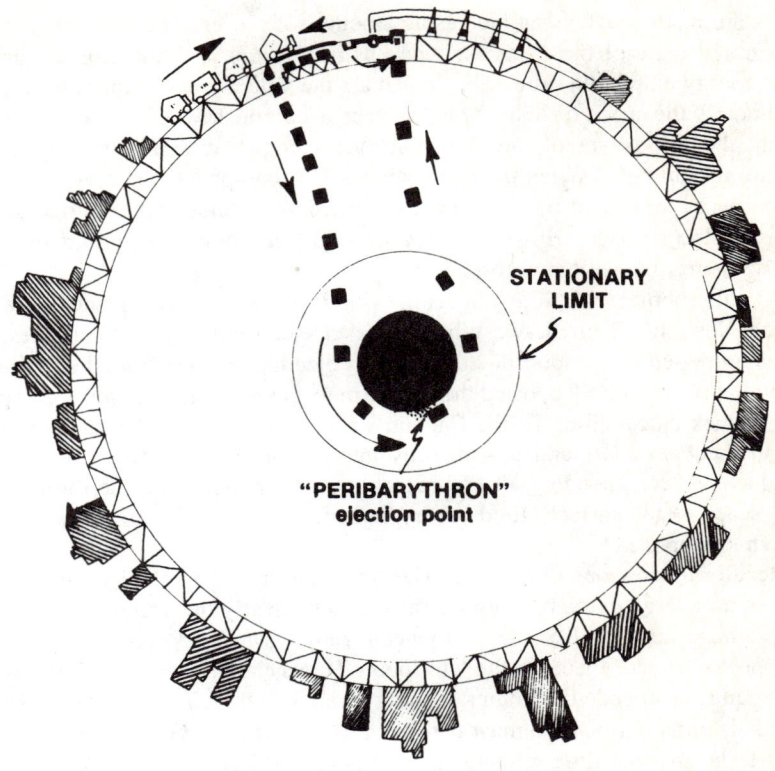

FIGURE 4. The black hole energy extraction machine. Shuttles from the shell enter the ergosphere, jettison garbage onto an energy extraction orbit, and return to the surface with a gain in energy to power a giant generator. The additional energy comes from the energy content of the garbage and from the rotational energy of the black hole. In this process, the hole slows down.
—From GRAVITATION by Misner, Thorne and Wheeler. W.H. Freeman and Company • 1973.

less. An inexperienced space cadet attempts to eject a load of garbage onto an energy extraction orbit simply by throwing it out by hand. To our dismay, we find that the garbage only follows the shuttle along, very gradually drifting away (exactly like what happens with garbage jettisoned from a space capsule in Earth orbit). The bundle is certainly moving too slowly to be on an energy extraction orbit; this garbage would hardly brake a snail, let alone a black hole. Determined, the crew tries again, this time firing the garbage out of a cannon. Now the garbage vanishes into the distance, but when our sensors plot the trajectory, we find that the garbage is still not moving fast enough to be on an energy extraction orbit. Many such attempts are made, each using increasing amounts of power. They all fail. Finally, the frustrated crew of the *Penrosia* succeeds in shooting a thimbleful of garbage onto an energy extraction orbit. They calculate the velocity of the thimble and find it to be nearly the speed of light. This has been accomplished only by momentarily diverting the full power of the shuttle's reactor engine, just for the purpose of launching the thimble. When the energy balance is computed, the crew discovers that the energy generated by the reactor on board was almost equal to that gained by ejecting the garbage. However, they have gained some energy and tired but happy, prepare to leave the ergosphere.

At this moment, the Captain realizes he has made a fatal mistake: He has forgotten Newton's Third Law. When a rocket ejects fuel from its engines, the rocket is propelled in the opposite direction. By ejecting garbage from the *Penrosia*, the crew has inadvertently boosted the shuttle onto a new orbit. The ship's computer makes a quick calculation. To the Captain's horror, he realizes that the new orbit will lead the *Penrosia*—and us—directly into the black hole. The Captain guns his engines. After expending all the energy gained by launching the thimble, we barely escape to the surface, tired but unscathed.

What happened?

Recall our previous discussion. The "sufficiently high velocity" mentioned earlier for an energy extraction orbit turns out to be nearly the speed of light. That is, the garbage must be like the Red Queen, moving at nearly the speed of light with respect to other objects in the whirlpool. To accelerate an object to the speed of light requires stupendous amounts of energy which, in this case, must be generated on board. It turns out that we must convert a large part of the garbage into energy in order to boost what little remains to the velocity of light.

The second problem was to get the energy out. Most orbits within the ergosphere intersect the black hole. The *Penrosia* boosted herself onto one of these orbits, and to escape it required also a stupendous amount of energy. In most cases, anything gained by the ejection is lost in trying to escape. This second problem, as it turns out, can be overcome only by jettisoning the garbage exactly at the peribarythron of the orbit. Then one escapes to the surface with what little energy was initially gained by the ejection.

We have just seen that to get an appreciable amount of energy out of the hole requires that matter be converted on board the shuttle with essentially 100% efficiency. Then by ejecting this "energy beam" (photons), we get an additional 20% boost from the hole, for a grand total of 120% efficiency. Not bad; but since this requires almost 100% conversion efficiency to begin with, the Penrose process might not be worth all the trouble. It can, however, be used for a more efficient energy conversion process than is available on Earth. That is, it turns out we can use a modified Penrose process to extract energy from matter with up to 10 times efficiency of the ½% of hydrogen bombs, the most efficient process known at present. Unfortunately, we do not have space in this article to discuss such modifications, and a more detailed discussion will have to wait for another opportunity.

"I Expect to Pass Through This World But Once": Star Gates and Time Machines

Any interstellar empire or commercial consortium needs a means of rapid communication and transport. The smuggler Han Solo made the "jump into hyperspace" and emerged at his destination some time (12 parsecs!?) later. Space warps and star gates are a staple of science fiction.

Relativity, as already mentioned, describes gravity as a warping of space and time, and a black hole is the result of the strongest possible curvature. It is not surprising, then, that science fiction has latched onto black holes in an attempt to make space warps sound more plausible. To some extent, it is our own fault; in idealized situations, relativists have discovered the tantalizing possibility of a "star gate" lurking in black holes. Unfortunately, the situation has gotten out of hand, and almost everyone has chosen to ignore work started as far back as a decade by Penrose and Floyd, which shows that "star gates" cannot be realized in practice.

To discuss this problem, we will need to back up and fetch some concepts not yet introduced in this article. Relativity is, in a sense, a study of geometry, but not simply the ordinary Euclidean kind which we all learn in high school. For one thing, space and time have been combined into a 4-dimensional space-time. To pursue this point briefly, let us refer to Figure 5. Here, only two spatial directions are shown, x and y (east-west and north-south if you like) and the time direction, labeled by "ct." Time increases upward. From the explanations accompanying Figure 5, we distill four rules for understanding these diagrams: 1) An object stationary in space still moves through time. Its path through spacetime, or *worldline*, is therefore a vertical line; 2) An ordinary, moving object, like a rocket, has a worldline which is tilted at less than 45 degrees from the vertical; 3) Light travels along 45-degree lines; 4) Traveling on a worldline tilted greater than 45 degrees from the vertical is prohibited because this is motion faster than the speed of light.

This type of spacetime diagram has its defects. The most serious one is the difficulty of showing things which are very far apart—it is especially difficult to map an infinite universe onto a finite piece of paper. Nonetheless, with sufficiently vigorous squeezing, one can actually distort the outer edges of the universe in such a way that we can fit the entire infinite universe onto a finite piece of paper. We can even retain certain features of the real universe. The one feature which is usually kept is the 45-degree angle which represents the trajectory of a light beam. A diagram like Figure 6 results.

All this is in preparation for a discussion of star gates and time machines based on black holes. Recall our discussion of tidal forces. We mentioned that in a simple Schwarzschild black hole, tidal forces on an infalling object (remember Richard's plunge) become greater and greater until they become infinite at the singularity. Well, inside a rotating Kerr hole, or a charged Reissner-Nordstrom black hole, this is not necessarily true. In fact, the theory states the following: *Unless the black hole has exactly zero charge and zero rotation, it will allow an*

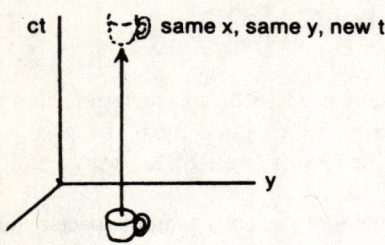
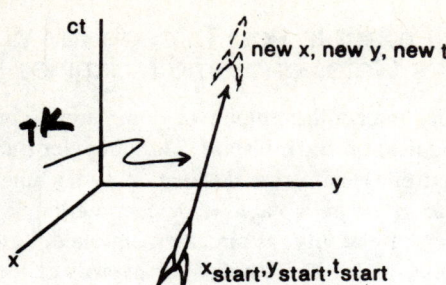

a) A stationary object in space still moves forward in time.

b) A moving object has a tilted world line.

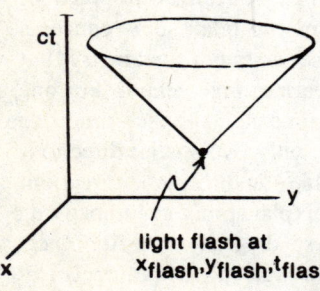
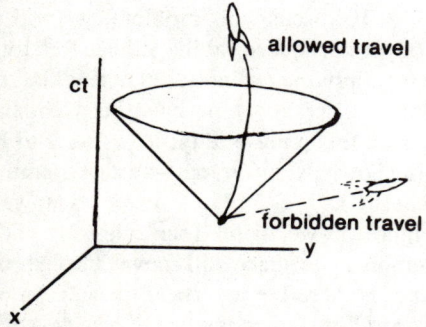

c) Light spreads out from a flash in all directions at 45° angles, forming a "light cone."

d) Since no physical object can move faster than light, any rocket must remain in its light cone; that is, the rocket must have a world line tilted at less than 45°. A tilt greater than 45° is motion faster than light and is forbidden.

FIGURE 5. Basic spacetime diagrams. The time scale is plotted vertically. The label "ct" is just to get the units correct.

object, say a spaceship, whose own gravitational effects are negligible, to enter the black hole at a speed slower than light, avoid the singularity, and leave again by an exit different from the surface through which it entered. The different exit surface gets around the immediate objection that nothing which falls into a black hole can escape again; it can, but not through the same surface through which it entered.

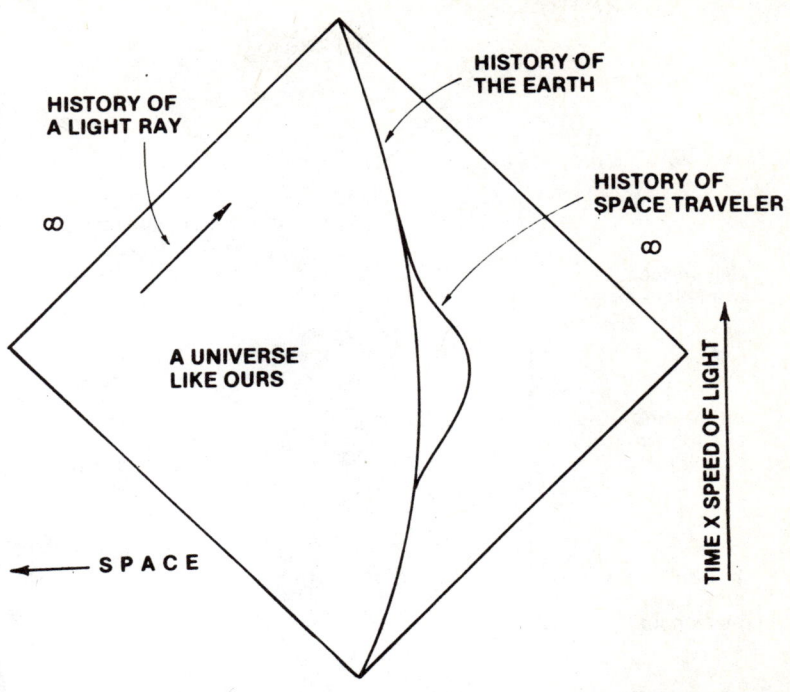

FIGURE 6. A history of our universe (much compressed around the edges). The space traveller's history looks longer, but according to the curious geometry of spacetime, actually amounts to a shorter time than passes on Earth between his departure and return.

If the exit surface is not the same as the event horizon by which we entered, then where does the spaceship end up? Figure 7, a spacetime diagram drawn for a Reissner-Nordstrom black hole, attempts to explain the situation. The two diamond shaped regions are like Figure 6, and thus both represent infinite universes. The diagram therefore shows two universes connected by a black hole tunnel. (Again, we are plotting space horizontally and time vertically.) The curve shows the history of the infalling-spaceship-cum-observer as it travels through the hole.

A complicated figure indeed. The intrepid traveller falls inward. The first 45-degree line he crosses is the event horizon of the black hole; nothing can re-exit via this surface once having crossed it. The jagged lines represent the singularities, where tidal forces are infinite. (Remember, singularities are moving forward through time; therefore, on this diagram they appear as vertical lines.) But the traveller can avoid these reefs; just by coasting he passes at a safe distance. When he emerges from the black hole, he finds himself in a normal universe like the one he just left. In fact, it may be precisely the one he left, but the black hole exit need not be near the entrance, and there is reason to think it would not be near that entrance.

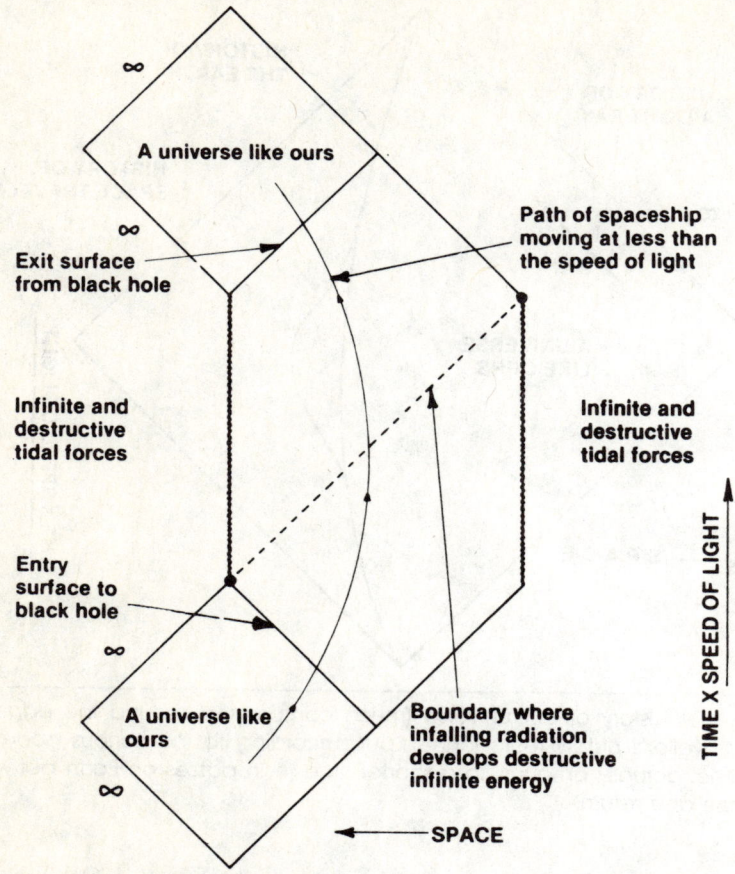

FIGURE 7. A spacecraft might fall into a black hole and hope to emerge via the star gate into a universe like ours—maybe in fact ours. (See Figure 3.) The dotted line is where infalling radiation can become destructively large and close this star gate, according to the General Theory of Relativity.

We might at first worry whether the second universe is the same as ours. According to the theory, there is no reason it should not be, but equally no reason it should be, either. (Parallel worlds!) If many charged or rotating black holes inhabited our universe, the possibility would exist that their exits would emerge in our own universe, or all in the same second universe, or in any number of alternate worlds (Figure 8). More bizarrely, these various universes might be connected in such a way that, after traveling through several black holes, one returns to our own universe, but at a previous time. The theory does allow for this possibility, suggesting the use of black holes as time machines. In any case, the first pioneer

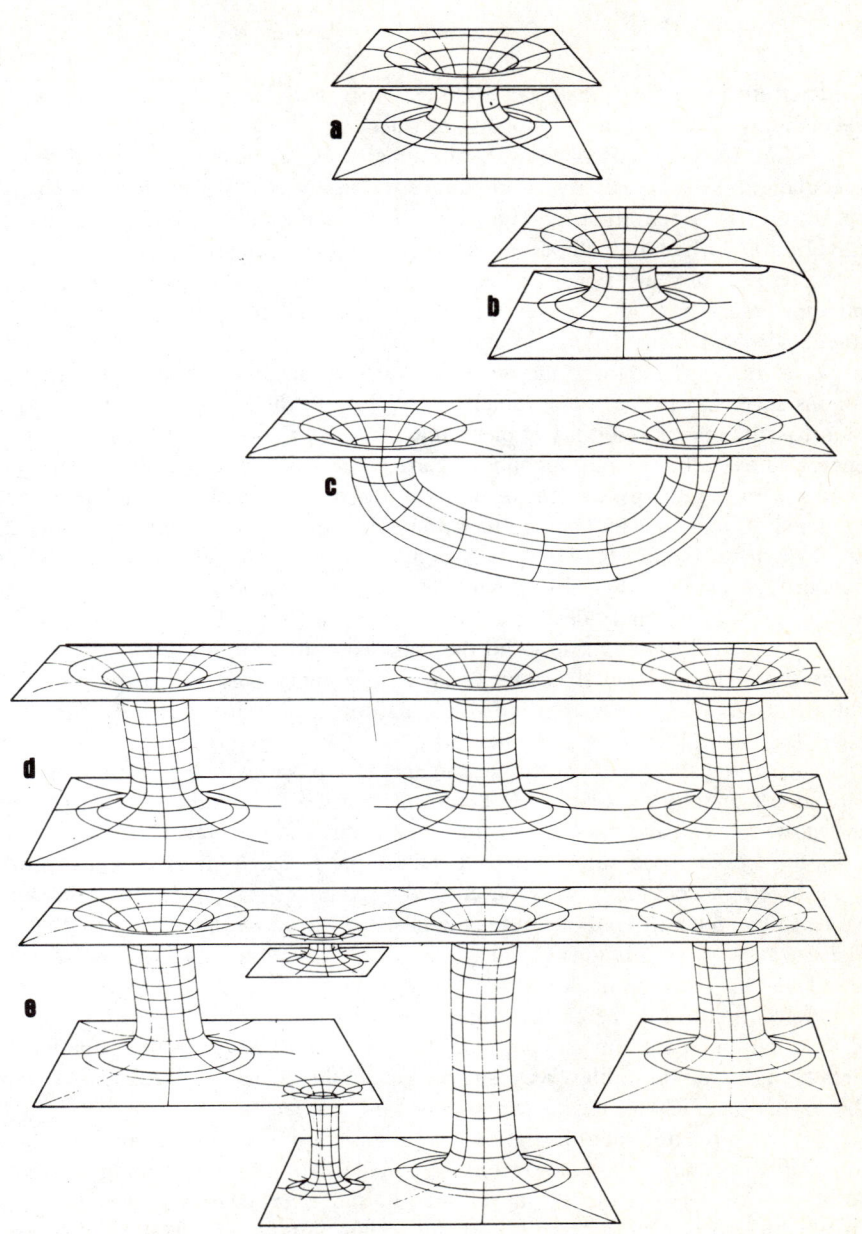

FIGURE 8. a) A schematic picture of a black hole connecting two disjoint universes; b), c) The two universes may in fact be the same; then the distance between the entrance and exit *might* be much shorter by the hole. d) If the two universes in a) are different, we would hope that at least every connection would be between the same two universes. Appealing to Murphy's law, e) is much more likely: a completely tangled web of interconnections.

to determine which of these possibilities exists will be a very brave man, and exceedingly dedicated to the progress of science.

Unfortunately, at this point, hard science drags us back from such interesting speculation. The crucial flaw in the above discussion was the assumption that the spaceship had a negligible gravitational effect on the hole. Can a real spaceship travel through the hole without disturbing its structure? In short, the answer is no.

To see this, just consider the energy content of solid space garbage, laser photons, radio and other waves, all of which a well-equipped interstellar space traveller would likely spread around during his trip. The central problem is that some of this stuff falls into the black hole too. As it falls into the hole, garbage, for instance, has picked up a velocity very close to the speed of light. We know that mass increases to infinity at these velocities. By $E = mc^2$, this means that the energy of even the smallest amount of garbage has also become infinite. The same occurs with the energy content of infalling radio waves and light signals emitted by the ship; their energy goes up to infinity also. But what has infinite matter and energy densities in a black hole? The singularity, of course. Where the clean black hole provided clear sailing, by allowing for garbage or radio waves we have created a singularity of infinitely destructive tidal forces, as in the Schwarzschild case.

A moment's thought leads to the conclusion that the black hole star gate is a one-time affair at best. If just the radio signals transmitted by the space traveller can be so disruptive, the mass of the spaceship itself must certainly disrupt the black hole and close the gate behind him.

Assuming he makes it through in the first place. What if he is extremely careful not to sully the black hole before his journey? He maintains radio silence and stows his garbage bags. Could he make it through the tunnel?

Surely, his spaceship must have substantial mass. A mass falling toward a black hole is accelerating and, consequently, generates *gravitational waves*, analogous to the generation of electromagnetic waves by accelerating electric charges. These waves are oscillations in the gravitational field which travel at the speed of light. Again, this is in analogy to radio waves which travel at the speed of light and are oscillations in the electromagnetic field. Some of these waves travel ahead of the infalling spacecraft and are amplified to infinite energy in the same way as already discussed for radio waves; and the gate to the other side of the galaxy slams shut in his face. The star gate cannot even be used once.

The simple argument just given shows that the inner structure of black holes is unstable to small disturbances from the outside. Any amount of energy or debris falling in from the outside will develop an infinite energy density and destroy the inner structure of the hole. In a perfectly clean universe, we could not know *a priori* whether a given Kerr or Reissner-Nordstrom black hole contains a tunnel to another part of the universe. But we do know that, if we probe the black hole by trying to reflect radiation off it, we automatically destroy the star gate, because some absorption of radiation is inevitable. In the real universe, of course, the situation is even worse because radiation and matter are everywhere present to some degree and must be falling into any existing black holes.

So it seems, for instance, the collapsar transport system used in Joe Haldeman's *Forever War* will not work; nor does the physics at the end of the Disney film hold water (not to mention the metaphysics); and, in fact, the syphoning of extra matter from another universe mentioned at the conclusion of T.R.'s own novel will not go through either. At least these works were released as science fiction. Books such as Adrian Berry's *The Iron Sun*, which purport to be science, are either flagrant rip-offs or bad science fiction. As either rip-offs or science fiction, they deserve no further serious consideration.

BIG SURPRISES COME IN SMALL PACKAGES: PRIMORDIAL BLACK HOLES

Until now, we have concentrated our attention on large black holes, from 10 solar masses to over 10^{10} solar masses. In this section, we turn our attention to the other end of the spectrum: mini black holes. At the very beginning of the universe, at times much less than one second after the Big Bang, the density of matter was comparable to what is found in typical black holes. It is conceivable, then—but by no means proven—that a slight fluctuation in density would "snap" the matter into black holes. Such "primordial" black holes would range in size from the very large, about 100,000 solar masses, to the very small, about 10^{-5} grams. The small holes would be formed first, when the density was highest, followed by succesively larger holes, until the density was too low to form any at all.

Large primordial black holes would behave in exactly the same way as the other large holes which we have already discussed. There is nothing to be added here. At the other extreme, holes of 10^{15} grams and below are remarkable objects. The density of 10^{15}-gram black holes is so high that one cubic centimeter of them would contain the known mass of the universe! Holes smaller than this mass would exhibit extraordinary quantum properties, specifically the famous Hawking radiation named after its discoverer. Space does not permit us to discuss these amazing properties. Suffice to say, there is no observational evidence to indicate that holes smaller than 10^{15} grams exist or existed. Moreover, theoretical upper limits placed on such holes by Page, Hawking, Novikov *et al.*, and two of us (R.M. and T.R.), indicate that, if they ever existed, they were few and far between. For instance, there cannot now be more than about 10 black holes of 10^{15} grams per cubic parsec, each with the mass of a mountain but the size of a proton.

Primordial black holes with masses greater than 10^{15} grams have negligible quantum properties and can be treated classically. Such black holes have also received attention in science fiction and popular folklore and therefore their share of misrepresentation. Perhaps the most famous—or notorious—suggestion was put forth by Al Jackson and Mike Ryan, then at the University of Texas, that the 1908 Tunguska blast in Siberia was caused by the collision of a 10^{21} gram black hole with the Earth.

We may first ask, "What are the odds of such a collision taking place?" Not bloody likely. Assuming all the observable mass in the universe to be concentrated into 10^{21}-gram black holes, one can calculate that one collision should occur about every ten ages of the Universe, or 10^{11} years. Marauding black holes do not seem an overwhelming threat to U.S. security. Nonetheless, it is possible that Tunguska was *the* collision. Although a 10^{21}-gram black hole is small in radius, about 10^{-7} centimeters, its mass is large, about one million small mountains. Jackson and Ryan proposed that the gravitational attraction of this hole caused the surrounding air to be yanked inward, resulting in a compact ball of air whose shock effects produced the destruction seen in well-known photographs. There was, however, substantial debate on whether a black hole of this mass would have the claimed effect when interacting with the solid earth. Most physicists believe the ground shock would have been tremendous, much more so than what actually occurred. So in scientific circles the matter is considered dead and buried. In any case, Al Jackson and Mike Ryan have on occasion confided that the suggestion was not entirely serious in the first place.*

Detailed statements about the interaction of smaller black holes (about 10^{15} grams) with matter are difficult to make. Nonetheless, simple calculations give the following general picture, which should not be too far wrong: Recall, a 10^{15}-gram black hole has the diameter of a proton. This is too small a size to rapidly accrete (gobble up) surrounding matter. Even if one considers that any nearby particle in random motion falls in when nearing the hole, one finds an accretion rate such that the black hole will not even double its mass in the lifetime of the universe. Talk of eating a planet becomes absurd. Thus, the black hole posited by Larry Niven in "The Hole Man" would certainly never gobble up Mars in less than extreme cosmological times, meaning millions or billions of ages of the universe.

WHERE DO WE GO FROM HERE?: PROSPECTS FOR THE FUTURE

We have talked about many types of black holes and many properties but have omitted discussion of many other interesting properties as well. We have not spoken about Hawking radiation, nor about black hole collisions, nor about superradiance, nor about astrophysical accretion, nor about photon trajectories and imaging properties, nor about the influence of primordial black holes on nucleosynthesis after the Big Bang. We have also shied away from direct discussion of the famous singularity which occurs within all black holes. The singularity, as already mentioned, is the center of the black hole where all the matter has fallen. It is a place where the density of matter is infinite, as well as gravitational and tidal forces. When people speak of space and time ending at black holes, they are perhaps thinking of the singularity. But it may be a mistake to say space and time end at the singularity; what ends is our present knowledge of physics.

---

*We have recently learned that a report in *Sotsialisticheskaya Industriya*, 1/24/80, indicates that ordinary meteoric debris was recently found at the Tunguska site.

Much work is currently under way to remedy the situation. Many physicists believe that true singularities do not exist, that at such small distances the quantum properties of spacetime itself come into play. They suggest that matter cannot be compressed to a smaller size than the so-called Planck length, about 10^{-33} centimeters, where the quantum effects become dominant. According to this view, the singularities of classical black holes are nonexistent in reality and are only the temporary nuisances of defective mathematics.

At least two Russian physicists, Frolov and Vilkovisky, have recently claimed to have proven that black holes, in some sense, do not exist at all. Proper use of quantum field theory, they argue, shows that as matter collapses to form a black hole, it misses the singularity, "rebounds," and eventually re-expands beyond the event horizon. This process, for even Tunguska-sized black holes, will take longer than the age of the universe. Nonetheless, in a strictly logical sense, a black hole is no longer a black hole, but only temporarily out of sight. We are not yet sure whether Frolov and Vilkovisky are correct, but we are certain that the full merger of relativity and quantum theory will reveal many answers and even more questions.

Milton A. Rothman

DEATH RISK

AUGUST 1980

WHEREVER we may go in the universe, we will face risks. Even if you never leave the chair in which you are reading this, there are influences at work that could conceivably kill you. Some, of course, are highly unlikely, such as a herd of elephants from a circus five miles away finding their way to your house and stampeding through your living room. Others, such as the danger of fire if your home is full of defective wiring, may be much more immediate concerns. How do you decide which risks are worth worrying about or taking preventive measures against? Both individuals and societies are constantly making decisions of that sort—but not always in the most rational way, as the following article shows.

Milton A. Rothman received his Ph.D. in physics from the University of Pennsylvania in 1952. He has worked in nuclear physics at the Bartol Research Foundation and in fusion research at the Princeton Plasma Physics Laboratory, and he taught at Trenton State College until retiring. He has written a number of science fiction stories, science articles, and books (most recently *A Physicist's Guide to Skepticism*, Prometheus Books, 1988), as well as research papers. He is also, by a remarkable coincidence, the father of Tony Rothman, coauthor of another article in this book. ∎

THERE are few things I know for sure. One of them is this: if I don't die of one cause, I certainly am going to die of another.

This being the case, why all the fuss about death risks? If you don't die of cancer, you will probably die of heart disease. Then why worry about radiation hazards? Why bother to clean up the environment? What difference does it make if a nuclear accident causes a thousand cancer deaths? A lot of those people would have died of cancer with or without a nuclear accident. And if there is an accident, be comforted by the fact that all the victims would have died anyway, sooner or later.

This peculiar state of affairs is known among demographers as the Taeuber Paradox. It doesn't take long to figure out the fallacy in the above paragraph. The giveaway is in the expression *sooner or later*. The important factor we have left out of the logic is *time*. Some causes of death strike before others. Therefore if a cause of early death is eliminated, you live longer, and if this is done for the entire population, then the average lifetime of the population will be increased.

Because of the time factor, a lot of the figures bandied about concerning radiation risks don't mean a great deal. It is not too informative to be told that so many millirem of radiation will produce so many cancers, because it takes 20 to 30 years for these cancers to germinate, and by that time the people involved might already be dead of other causes.

It is more informative to know how a particular risk changes the average lifetime of the population if that risk is sustained over a long period of time. If we know that risk A reduces the average lifetime by two years, while risk B reduces the average lifetime by one year, then we can say that risk A is twice as bad as risk B. It makes a neat way of comparing risks.

Another way of looking at it is to ask how much the average lifetime would change if a particular risk were removed. In that event we would expect an increase in the average lifetime. For example, in the year 1900 the average lifetime in the United States was only 48 years. (Statisticians call this number the *life expectancy*, because if people on the average live to be 48, this is how long you expect to live. In reality, of course, some people live less and some people live longer than average.) By 1975 the life expectancy had risen to over 72 years, largely as a result of improved public health and the eradication of contagious childhood diseases such as smallpox, scarlet fever, polio, etc. (Anybody who thinks the health professions haven't done anything for us ought to contemplate these figures.)

You see that elimination of certain early risks has dramatically increased the mean lifetime. One consequence of this situation is that a large number of people now live long enough to get the cancers and heart diseases that they didn't get in the good old days, simply because they didn't live long enough. Another consequence is that since people now *expect* to live to be at least 72, they are much more upset than they used to be over the prospect of dying at 50 or 60. A 60-year-old man does not seem old nowadays. (I am just 60.)

For this reason, environmental risks are taken more seriously now than they were previously. Now that science has increased the average lifetime from 48 to 72, the public resents the possibility that any of the by-products of science might take a little bit away from those extra years that have been gained. This is a natural reaction, but too often judgments about risks are made on a basis other than reality. (For example, small amounts of radiation are feared more than large amounts of cigarette smoke or coal dust.)

In order to make valid judgments about risks, we must be able to make numerical comparisons between various competing risks. We would like to ask:

how much will nuclear power affect the average lifetime compared with coal power? However, before we do that, we should know what is the present risk of dying due to all causes. Then it is easy to see what happens when one particular risk is increased or decreased. We can get this information from statistics published by the U.S. Department of Health, Education, and Welfare. From their 1975 Life Tables much of the data for this article was obtained.

What these Life Tables do is to assume a group of 100,000 people (a cohort) born at a given time. Then they suppose that during their lifetime these people are exposed to the same risks that prevailed during a particular year—in this case 1975. The tables then show how many people out of the original group of 100,000 will still be alive after a given number of years have elapsed. The results can be put into the form of a graph called a survivor's curve (Figure 1). (This particular curve is for the total population. Statistics are available for males, females, whites, and non-whites. The various groups show important differences: women live 7.8 years longer than men, on the average, and whites live 5.3 years longer than non-whites, on the average.)

The important features of the curve are plain to see. There is a brief drop of about 2% during the first year or two, there is an almost flat period till about age 30, and then the decline gradually steepens until by age 76 half of the original group is gone. (Note: this half-life of 76 years is not the same as the average lifetime of 72.5 years.) Beyond age 90 there are few left.

We can now go ahead to answer many interesting questions, such as: out of those still alive at the beginning of a given year, how many will die during that

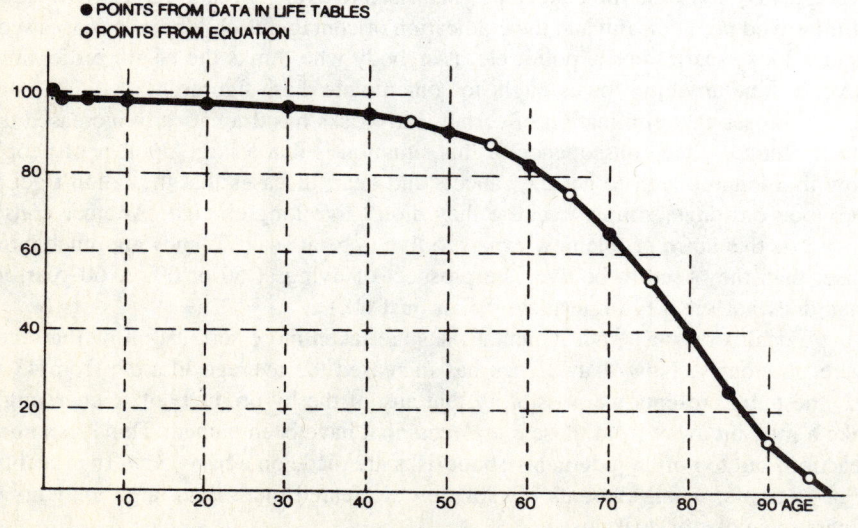

FIGURE 1. Graph of the percentage surviving past a given age out of a group born during a single year. (1975)

year? Even more interesting: what *fraction* of those alive at the beginning of a year will die during that year?

For example, we find that at age 40 there are 94,226 survivors, and at age 41 there are 93,973 survivors. This means that 253 people died during that year. Dividing 253 by 94,226 we find that 0.00269 (0.269%) of the people reaching the age of 40 died during the following year. But this number is exactly what the statisticians mean by a probability. The numbers tell us that the probability of dying between the ages of 40 and 41 are 0.00269 or 2.69 chances out of a thousand.

Proceeding further, we find that out of the 65,529 surviving to age 70, 2,176 (or 3.32%) will die during the next year. Going from age 40 to age 70, the probability of dying increases more than tenfold to 33.2 out of a thousand. There is nothing startling about this; the older you get, the greater your chances of dying in any given year.

What we would like to know is exactly how this probability increases with time. What is the mathematical relationship? To get a handle on this, we plot a graph that shows the probability of dying during a given year, using the data in the life tables (Figure 2). Study of this graph shows a number of important features. First of all, there is the dip during the first few years. This dip arises from genetic defects and sundry ailments that very small children are prone to. Once they make it past the age of 5 they are on relatively safe ground. Then, past the age of 15 the curve rises again and becomes a strikingly straight line. (A comparison of this curve with those from the earlier part of the century, or with those from underdeveloped countries, shows the low portion of the curve lying much higher than the U.S. 1975 curve. The effect of medicine and public health has been to depress the middle, low-lying section of the curve.)

Now let's look carefully at the straight line that we find between ages 30 to 80. Notice first of all that the vertical scale is a logarithmic scale. Going up by equal intervals on the graph means that the number is multiplied by a constant factor. Each main division on the scale means a factor of 10 increase in probability.

A straight line on a semilogarithmic plot like this has but one meaning to any mathematically trained mind. The probability of dying increases *exponentially* as you grow older.

Exponential increase is the kind of increase you get from compound interest, or from any kind of growth that takes place at a constant annual rate. If you invest money at an 8% interest rate (compounded continuously), in 8.6 years your money will double. In another 8.6 years, it will double again. And so on.

That is exactly how the death curve rises. The probability of dying increases at an 8% annual rate, and doubles every 8.6 years. What happens beyond age 80? The tables merely give the probability for the group 85 and over. This number is 1.0000.

Alle Menschen müssen sterben.

Generally speaking, rising exponentials like this one come from specific reasons, usually related to positive feedback. As far as I can tell, and I have asked some people who ought to know, there is no generally accepted explanation for

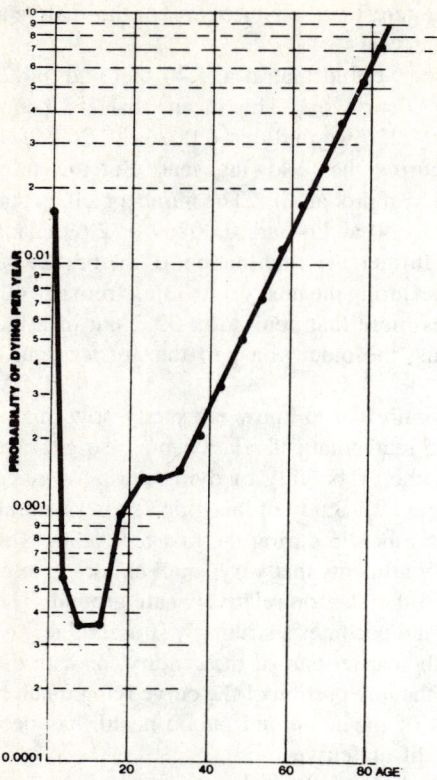

FIGURE 2. The probability of dying in a given year, for a group born during a single year.

the life curve. We can speculate about the growth of mutated cells within the body, or the decay of immune mechanisms, or a combination of the two. Clearly, this is a major topic of research at the present time.

Regardless of the cause, the nature of this curve hints to us that there is a rock-bottom level of mortality that is built into human physiology, and that after you subtract out the effects of bacteria, viruses, and other environmental factors, the cells themselves gradually wear out. In other words, the body is not designed to last much more than about 110 years. We would expect that efforts to increase this basic longevity will have to aim at the fundamental molecular processes, and some scientists believe that cellular aging can indeed be slowed down.

Once we know that the probability of dying is represented by an exponential curve, it is a very simple matter to write down the equation that represents the curve. In doing this we ignore the bottom part of the curve and use only the straight part. This will make little error in any calculations we perform with the equation, because relatively few deaths occur before age 30.

Now we have a mathematical equation from which we can calculate the probability of dying at any given age (past 30). (See the Appendix for details.) From this equation we can derive another equation—an equation that tells us how many people will survive to a given age. But that's just the information we plotted in Figure 1. So now we have equations that fit both curves, and with these equations we can calculate such things as average lifetimes, the probability of surviving to a given age, and—most important—what happens to the average lifetime if certain risks are varied by a given amount.

Since the mathematics does become rather sophisticated, we can save ourselves a great deal of trouble by using the graphs to draw some conclusions, instead of working with the equations. To demonstrate, let us ask the following question: suppose a catastrophe occurs that causes the probability of dying each year to double. How does this affect the average lifetime?

To answer this question, we simply double the value of every point of the straight line of Figure 2. The result is that this line slides straight upward without changing direction (Figure 3). (Remember that when you multiply a number by 2, you just add 0.3 to the logarithm of that number, so doubling just raises each point on the curve by the same value.) Now notice that the same result could be obtained by sliding the straight line to the left a distance of 8.6 years.

In other words, the new curve, with double probability of death, is just like the old curve, but with everything happening 8.6 years sooner. The result is that doubling the death probability reduces the population life expectancy by 8.6 years. (This argument glosses over some details, but the results can be verified by using the equations in the Appendix to calculate the new life expectancy and seeing that, to a good approximation, the simple argument given above is correct.)

Now this sounds like a paradoxical result. You would think that doubling the death risk would cut the average lifetime in half—from 72.5 years to about 36 years. Instead, the life expectancy is reduced to about 64 years. The catastrophe is not catastrophic, as we might have expected naively. The reason for this result lies in the steady rise of death risk that goes on all the time. If right now something happens that doubles your risk of dying, it simply puts you in the same place that you would have reached normally, 8.6 years from now.

Well, 8.6 years may or may not sound like a big change; it depends on your point of view. So here is another example to try out for size: what would happen if all cancer was eradicated from the population? How would this affect the average lifetime?

Looking in the appropriate tables (*Health, United States, 1976–1977*, U.S. Department of Health, Education, and Welfare), we find that cancer causes about 28% of the deaths in those above 40, and the percentage is roughly constant with age. This means that if we remove cancer as a cause of death, we must lower our probability line by a factor of 28%.

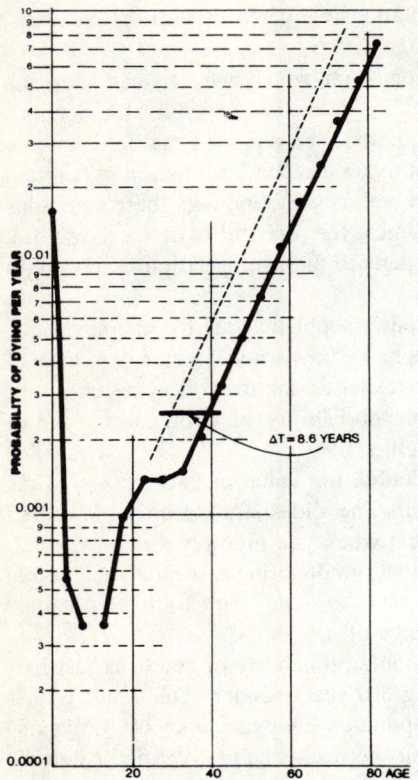 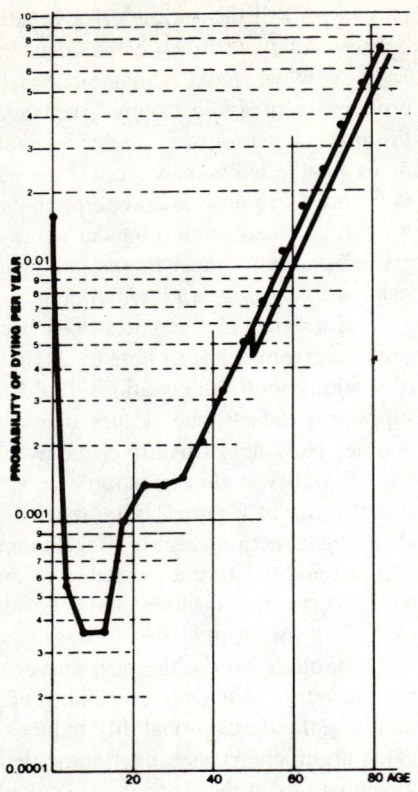

FIGURE 3. The probability of dying per year is doubled. This change is equivalent to shifting the line to the left by 8.6 years.

FIGURE 4. If the probability of dying decreases by 28%, this is equivalent to shifting the probability line to the right by 4 years. If the change suddenly occurs when the cohort is at age 50, then at age 54 the probability of dying is just as great as it was before the change took place.

This operation results in moving the line to the right by approximately 4 years. Repeat: *as a consequence of getting rid of all cancer, the life expectancy of the U.S. population is increased by only 4 years*. (In reality, less than 4 years, since people prone to cancer are also prone to other diseases.)

It would seem that the mountain has labored to bring forth a mouse. All of the billions in cancer research going into an effort that will increase the population lifetime by a mere 4 years! Why is this?

Figure 4 explains why this is. Suppose we could suddenly, miraculously, get rid of cancer at one point in time. The probability curve would suddenly drop to a lower level. However, all the other causes of death would continue to increase exponentially in their usual way, and after 4 years the probability of dying would be right back where it was before—just delayed a little bit.

Of course, we must realize that the 4-year figure is just a bare statistic and doesn't tell the whole story. Partial truths are almost as bad as lies. First, a 4-year increase in life expectancy is a big, big effect. It means a lot of extra man-years for the population as a whole. Now that infectious diseases have been conquered to a large extent, further increases in mean lifetime are hard to come by. Every extra year from here on out is going to take a big effort. Second, some individuals gain much more than 4 years through the cure of cancer. Remember, the 4 years is only an average. Third, cancer is a nasty, painful, and expensive way to go. So even if you merely replace cancer with heart failure, that's some kind of improvement.

Now getting rid of all cancer must still be considered a fantasy for the future. However, there is one improvement that *could* be made right now if everybody *wanted* to hard enough. And that is for everybody who smokes to give it up. The risks of tobacco smoking have been publicized enough, but they are put into a new light by a study recently reported in the *New England Journal of Medicine*. A group of 4,000 smokers was followed for 11 years, and the death rate for this group was compared to that of a similar group of non-smokers. (Ages were between 35 and 54 at the beginning of the study.) Notice that it was the *total* death rate—from all causes—that was being compared. Not just lung cancer or heart disease. The death rate among the smokers was 9.02 per 1,000 person-years, and 3.54 per 1,000 person-years among the non-smokers. In other words, the smokers had a death rate 2.6 times greater than non-smokers during the period of time. Their risk of death more than doubled!

In terms of life expectancies, these numbers mean that the smokers have 10 years less average lifetime than the non-smokers. The effect on the total population depends on how many people smoke. Recent figures indicate that about 33% of the population smokes. We would expect the overall life expectancy to be reduced by 3.3 years as a result. Any way you look at it, this is a big, big effect. And it is an environmental factor that can be turned on and off at will. It is a much bigger factor than many environmental hazards that commonly arouse national hysteria. (It would be interesting to know how many anti-nuclear demonstrators smoke. The statistics are not yet in on pot, but from general principles I would not like to have any kind of smoke in intimate contact with the cells of my lungs.)

With the growing clamor concerning nuclear energy, the only way to put a sensible perspective on the situation is to perform the same kind of analysis concerning radiation risks as we have done for cigarette smoking. I don't like the kind of glib argument that says: if we have X number of nuclear reactors, then there will be Y number of cancers as a result. It makes more sense to say: if we have X number of nuclear reactors, then this will produce a certain change in the probability of death each year, and the result will be a loss of N years from the life expectancy of the population. This gives us a number that can be compared meaningfully with the risks from other methods of producing energy.

It happens to be a difficult kind of analysis to do, because you must know in detail how a given amount of radiation affects the death rate over a period of time. However, we can make a rough stab at it and hope that our results are not wrong by more than a factor of 2 or 3. Which, under the circumstances, is close enough.

We create a scenario like this: the maximum level of radiation allowed at the boundary of a nuclear power plant is 5 millirem per year. It is estimated that if nuclear power capacity increases to 800,000 megawatts by the year 2,000, then the exposure to the general population will average out to something like 1 millirem per year per person.

We then go to the chart that says the cancer risk from radiation is about 0.0003 cancers per person per rem exposure. Multiplying this number by 0.001 rem per year, we find that the probability of developing cancer from nuclear power under the conditions given is about 3×10^{-7} per person per year. Comparing this number with the present cancer death rate (and assuming that each additional cancer will result in a death, which is overestimating the risk) we find that the nuclear radiation increases the probability of dying of cancer by 0.005%. This raises the probability curve by an amount too small to read on the graph.

Since we can't find the change in life expectancy from the graph, we must go to the equations. Application of a little calculus to the first equation in the Appendix gives us the following formula: the change in mean lifetime (in years) is equal to the percentage change in probability of death due to the added risk (0.005%), divided by the annual percent increase in the probability of death due to all causes (8%). Putting in the numbers: 0.005/8 equals 0.0006 years, which comes to about 5 hours.

In other words, the assumed increase in risk produces a loss of 5 hours out of about 72 years life expectancy—a change of about 8 parts per million. Is this worth worrying about? Well, that depends on your point of view. It depends on the kinds of things you like to worry about. (And that's why discussions of these matters are often so futile—different people worry about different kinds of risks.)

For example, an increase of 8 parts in a million results in approximately 27 additional deaths per year out of the whole U.S. population. That makes nearly 2,000 deaths over a lifetime.

Headline: NUKES KILL 2,000 PEOPLE.

Sounds bad, doesn't it? But hold on a minute. Right now, medical diagnostic X-rays are zapping the public with a dose estimated to be 72 millirem per person per year, on the average. That's 72 times greater than the figure we gave for all the nuclear power plants projected for the year 2000. Using the same arithmetic, we could just as well write a headline: MEDICAL X-RAYS KILL 144,000 PEOPLE.

That's called lying with statistics.

After all, the medical X-rays are needed to prevent deaths by illness in the first place. The X-rays might cause 1940 deaths per year, but how many would have died *without* the X-rays? The idea is to prevent more deaths than you cause. Otherwise the business is not cost effective.

The argument could be made that the risks from X-rays differ from the nuclear power risks in that one risk is willingly assumed, while the other is imposed on a person when he takes a medical X-ray. Regardless: the risk is assumed willingly or unwillingly in order to obtain a benefit.

Furthermore, you can obtain the benefits of X-rays while reducing the risk simply by improving the equipment and photographic techniques, and by not x-raying people especially sensitive, such as fetuses in utero. By reducing the average X-ray dose by a few millirem per year, we can more than compensate for the additional dosage gotten from nuclear power.

Another way of looking at it is to notice that inhabitants of high places such as Denver get twice as much cosmic radiation as do sea-level people. The background radiation at sea level is about 100 millirem per year. Living in Denver adds another 100 millirem per year. So the radiation risk projected for nuclear power plants is only 1% of the *extra* risk that people in Denver already live under. Is this something to worry about?

Those who insist on worrying will worry. Behold the hysteria that prevailed at Three Mile Island, where the radiation dose due to accidental leakage was much less than the *extra* radiation that people in Denver live with normally.

You must understand that so far I have been discussing only the normal operating risks of nuclear power. Problems with truly catastrophic accidents, waste disposal, and the proliferation of fissionable material are a horse of another color. I am not fanatically disposed in favor of nuclear power, being a fusion man myself. My purpose is merely to warn against irrational and superstitious fear of radiation risk. The way to avoid such fears is to look at the numbers and see that normal radiation risks are much less serious than many risks we live with continually.

By now it is a banality to say that anything you do has some risk. It is not so commonly noted that even doing nothing creates risk. Suppose, for example, that we do nothing to increase the supply of energy to meet the demand. Then comes hot summer days when we have power blackouts. It is noticed that the death rate among the elderly goes up sharply on such days. Air conditioning is not a luxury for everybody. It can be as much a necessity as antibiotics.

Therefore, the only way to discuss risks intelligently is by comparing the risk of one course of action against another. Herbert Inhaber (a consultant for the Canadian Atomic Energy Control Board) makes this point in the February 23, 1979 issue of *Science* (the journal of the American Association for the Advancement of Science). Inhaber compares the risks associated with the production of power by all the usual methods, and surprisingly concludes that solar energy is even more risky than nuclear power.

This sounds like a most paradoxical statement, going against all preconceived notions. However, the way Inhaber arrives at this conclusion is to add up *all* the risks connected with a given power source: not just the risks associated with the operation of the plant, but the risks that go with the construction of the plant and digging the raw materials out of the ground. Solar power is very dilute, requires large structures, and so requires very large expenditures of material per kilowatt output. Therefore, the risks connected with digging the materials for the concrete, the copper, the steel, the silicon, the risks of the refineries, the transportation, the plant construction—all of these well-known occupational risks are relatively high for solar power. Even installing a solar heating panel on the roof of a house involves the risk of falling off the roof, or at least dropping something off the roof.

Inhaber has been roundly criticized for numerical errors in some details of his calculations. For that reason I am not going to quote his exact conclusions. Better calculations will surely be made in the future. Nevertheless, the basic idea remains true: in order to talk about risk properly, you must compare risks of different courses of action. And when you do that you must include all the risks, from the beginning to the end of the process. You must tell the whole truth, not just a part of it. Under these conditions solar power turns out to be not as completely benign as its enthusiasts make it out to be.

The risks of solar power are easily overlooked because they are hidden in the background. The consumer doesn't see them. All the risks are borne by the copper miners, the gravel-pit operators, the steel workers, the electrical workers, and all the others who make it possible for that clean electricity to flow from the consumer's outlet at the flick of a switch. To some extent this is true of fossil fuel power also. For many years the housewife was completely unaware of and uninterested in the coal miners who were the ultimate producers of the energy she used.

We have here an area of ethics which needs exploration. Does the consumer have a right to clean energy with no risk attached, while the coal miner and oil driller and nuclear worker operate under a high level of risk? The argument is usually made that high risk is allowable if the individual involved enters into that risk voluntarily (as when you get into your car to drive down the road), while risks imposed from the outside by industry and government are intolerable in any degree. I am not sure that this argument is completely tenable. Not all coal miners really want to be there.

On the other hand, it is simplistic to make the argument that, well, the caveman had risks from sabre-toothed tigers, and medieval man had risks from the plague, and since man has always had to live with risks, we should stop whining and live with our risks as best we can. (And get rid of EPA.) While it's true there has always been risk, and the average person lives with less risk now than ever before, that's what civilization is all about. We've gone to a lot of trouble to create this civilization where the average person can live better and longer than he could before. What we don't want to do is to backslide because of the excesses of this civilization. So risk assessment becomes a full-fledged academic specialty and the debate goes on furiously.

There are those who would call down a plague on both houses and turn away from modern civilization, expecting that a return to rural simplicity will solve all their problems. I imagine that a great many of these people are too young to remember what it was really like in that nostalgic era we imagine to have existed at the turn of the century. Just try to picture what it must have been like at a time when the life expectancy was only 48 years. Visualize the fantastic increase to the present 72 years, and understand what has been done to create that increase.

When the 20th century is reviewed, this increase in life span, with all its ramifications, may turn out to be the most important single development of the century.

Whatever happens in the future, I want the world to have enough of a technological base to support the manufacture of polio vaccines and computerized X-ray machines. I want to have enough power and mobility so that I can get quickly to a doctor when I need one, and so that I can have a reasonable amount of air conditioning.

In some ways the world has gone downhill during the past 50 years, but I am old enough to remember the terrors of diphtheria and scarlet fever, old enough to remember the young man just quietly keeling over at the desk next to me in a non-airconditioned Washington office with the thermometer hanging around 95°F. We've come a long way (all of us) and I don't want to give up the good things we've gained.

So I propose a new motto: ENERGY IS GOOD FOR YOUR HEALTH. And if you think having energy is too risky, just think about how risky it is not to have it.

APPENDIX

Consider a number of people born in a given year. We will call this number n_0. Let n be the number of people who survive past the time t. If n_0 is taken to be 100, then n will be the percentage of survivors at time t. Let dn be the number of people who die during the interval of time dt. The fraction dn/n represents the probability pdt of dying during the interval of time dt. From Figure 1 we see that this probability is a rising exponential, so we make the hypothesis that it can be written in the form

$$\frac{dn}{n} = -pdt = -ce^{at}dt \qquad (1)$$

(The minus sign arises because n is decreasing as t increases.)

Those of you familiar with the theory of radioactive decay will recognize the left side of the equation. In the radioactivity problem the probability of an atom disintegrating per unit time is a constant. Here p is an exponential function of the time. Constants c and a are numbers to be determined from the data.

Equation (1) may be integrated to find n, the number of people left alive at time t:

$$n = n_0 \exp[b(1 - e^{at})] \qquad (2)$$

where $b = c/a$. A very good fit to the Life Tables of 1975, using data for the whole population, can be obtained with the numbers $n_0 = 98.4$, $a = 0.0801$, $b = 0.00153$, $c = 0.000122$. (Using 98.4 for n_0, instead of 100, takes into account the small drop during the first year, which the above equation does not include.)

A comparison of the data points and the theoretical points of Figure 1 shows agreement as good as you will ever find in a graph.

The unusual nature of Equation (2) is apparent. It is an exponential of an exponential. For small values of t, the quantity in brackets stays close to zero and n decreases linearly. When t gets past 40, then the e^{at} term zooms up, and the quantity in brackets becomes a large negative number and rapidly drops toward zero.

REFERENCES

The Effects on Populations of Exposure to Low Levels of Ionizing Radiation, National Academy of Sciences, Washington, D.C., 1972.

Friedman, G.D., *et al.*, "Mortality in Middle-aged Smokers and Non-smokers," *New England Journal of Medicine*, February 1, 1979.

Health—United States—1976-1977, U.S. Department of Health, Education, and Welfare, Washington, D.C.

Inhaber, Herbert, "Risk with Energy from Conventional and Nonconventional Sources," *Science*, February 23, 1979.

Keyfitz, Nathan, "What Difference Would it Make if Cancer were Eradicated? An Examination of the Taeuber Paradox," *Demography*, November, 1977.

Morgan, Karl Z., "Cancer and Low Level Ionizing Radiation," *The Bulletin of the Atomic Scientists*, September, 1978.

Vital Statistics of the United States, 1975, Vol. II, Section 5, Life Tables, U.S. Department of Health, Education, and Welfare, Washington, D.C.

Richard Patrik Terra

HOT ROCKS

AND WATER

MARCH 1985

THIS article is an intriguing example of how seemingly unrelated sciences—in this case, oceanography, biology, geology, and astronomy—interact and help each other along. Deep-sea expeditions on Earth, seeking evidence pertinent to a geological theory only recently granted general acceptance, stumbled onto communities of living things in a place where everyone had assumed they could not exist. "Everyone knew" that all life depended on sunlight, but ecosystems around hydrothermal vents where no sunlight could reach clearly proved that other sources of energy could support life as well. Meanwhile, space probes were finding similar conditions, previously unsuspected, on other bodies of the Solar System. Putting the two discoveries together immediately and dramatically broadens the possibilities for extraterrestrial life.

Richard Patrik Terra was born and raised in central Idaho. He attended the University of Washington and holds degrees in biology and communications. Most of his published work has been factual science articles, but he also writes science fiction and fantasy and is a graduate of the Clarion West writer's workshop. He is currently employed as a protein purification technician with a Seattle biotechnology firm. ■

EVERYWHERE on the face of the Earth, we've found life of some kind—everywhere under the sun. Up until the last decade or so, in fact, it has generally been accepted that it is the energy of Sol's rays that drives the ubiquitous biological activity of this planet. It seems intuitively obvious that the energy locked up into organic compounds by photosynthetic organisms forms the foundation of the complex food-chain pyramids upon which the rest of terrestrial life depends.

In recent years, however, this "obvious" observation has been questioned. We've now begun to find some remarkable exceptions that prove such questions to be worth asking, and that raise others equally intriguing. But if life is not to be dependent on the light of the sun, either directly or indirectly, what other sources of energy might it turn to? What is the alternative to photosynthesis? It is now becoming apparent that one of the best alternative energy sources on this planet *is* the planet. It seems that all it takes are some hot rocks and liquid water.

The interior of the Earth is a tremendous heat engine. Perhaps as much as half of the heat energy deep inside the planet is a primordial remnant from the accretion, gravitational compaction, and differentiation of the early Earth into crust, mantle, and core. Over the lifetime of the planet, this residual heat has been augmented by the slow, steady decay of radioactive elements such as Uranium and Thorium. The Earth has been slowly cooling for about 4 billion years now, with most of the heat being carried toward the surface by convection currents in the solid, semi-plastic rocks of the mantle.

This vast geothermal energy is the driving force behind plate tectonics, the slow wandering of the giant jigsaw pieces of the Earth's crust. The continents ride atop these plates as they march across the globe, separating, colliding, riding over one another's margins. New crust is formed in the sea beds of the mid-ocean rifts, where drifting plates are slowly pulling apart. Fresh magma wells up from below into the rift, cooling and solidifying into broken fields of pillow lava and adding to the margins of the separating crustal plates.

Geologists and oceanographers are deeply interested in these deep-sea spreading centers, and have been studying them intensively for years. Such studies began in the late 1960s with international diving expeditions like Project FAMOUS in the mid-Atlantic, and later moved to the Caribbean. In the late 1970s the focus shifted to the East Pacific Rise, a vast system of undersea rifts and faults that runs along the west coasts of South and Central America.

There seems to be little apparent connection between the slow, majestic movement of tectonic plates and life—between *geological* and *biological* activity. When exploratory dives, sponsored by the National Science Foundation and conducted jointly by the University of Oregon and the Woods Hole Oceanographic Institution, began on the East Pacific Rise in 1977, the scientists involved certainly saw none; biologists were not even included among the expeditions' geologists and oceanographers.

The first expeditions in the Pacific were aimed at exploring the Galapagos Spreading Center, a region of the sea floor where two smaller plates of the Earth's crust bordering South and Central America are pulling away from each other, and also from the vastly larger Pacific Plate. Because of this slow triple separation, the region is of special interest in the study of sea-floor spreading. The dives were made at a site just a bit east of the Galapagos Islands on the equator, hence the name.

Preliminary probes were made by towing cameras and other instruments close to the sea floor, which lay between 2,500 and 3,000 meters below the surface. In the course of making these criss-cross runs above the rift, several temperature anomalies were noted, indicating plumes of slightly warmer water rising up amid the very cool seawater above the rift. The investigators realized they were probably seeing some sort of hydrothermal activity associated with the upwelling of magma into the rift—a sort of slow undersea geyser or hot springs—and were eager to send down their manned submersible (in this case, the *Alvin* from the Woods Hole Institute) for a closer look.

The existence of such hydrothermal vents had been hypothesized several years before in an effort to help explain some puzzling aspects of the mineral balance of the oceans, and to account for some of the trace components dissolved into their waters. Calculations had shown that the runoff of all of Earth's rivers, carrying their huge loads of silt and dissolved minerals, would give the oceans a greater concentration of magnesium and a lesser one of manganese than were actually observed. It was thought that as the cool ocean waters percolated down into the hot, newly formed crustal rocks at the mid-ocean rifts and were heated, they would drop off some of the excess magnesium and take on the needed manganese to produce the balance that is actually found.

The hot, geothermal waters would also take on large quantities of sulfur and its compounds, such as hydrogen sulfide (H_2S), sulfates and sulfides of iron, copper, zinc, and other metals, as well as trace gases—methane (CH_4), hydrogen, carbon monoxide, and nitrous oxide (N_2O).

The vents themselves were easy enough to find: the final outflow of the circulating geothermal waters emerged from openings in the fields of pillow lava on the sea floor approximately 15°C to 25°C warmer than the surrounding seawater, at a frigid 2°C. The warm waters flowed from the cracks and crevices at rates of several centimeters per second, and the gentle streams often had a milky blue-white color. Through the portholes of the *Alvin*, the investigators peered through the dark, near-freezing waters into the bright pools of illumination cast by their floodlights. What they saw was as fascinating as it was unexpected.

In the midst of the vast, rolling expanse of barren lava formations, far below the level where the last sparkle of sunlight might possibly penetrate, they found flourishing oases of life, thriving in seeming defiance of the cold, the darkness, and the crushing pressure, which at those depths is nearly 2 tons per square inch. Many of the vents were surrounded by lush communities of benthic (deep-water) organisms arranged in concentric, overlapping rings of life (Figure 1-A.).

The vent outflow mixed rapidly with the surrounding seawater, its temperature quickly falling from about 25°C to 3°C. Huddled nearest to the warmth were dense clusters of huge, long-bodied tubeworms, their fragile homes rising from the rocky sea floor like a white, skeletal forest. From the open ends of the tubes, the worms

100 / THE UNIVERSE WE LIVE IN

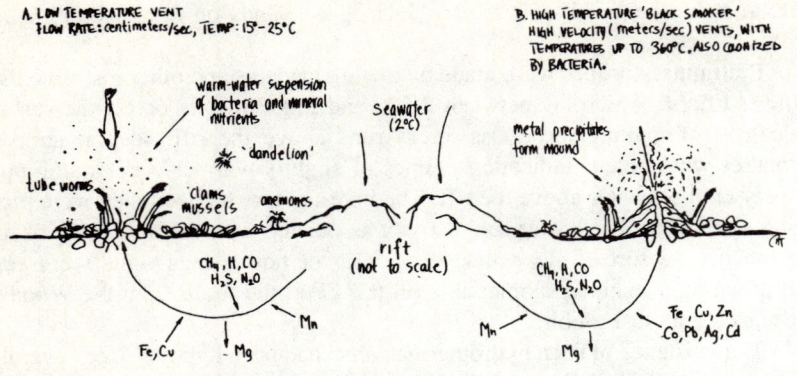

FIGURE 1. Diagram showing the two types of hydrothermal vents described in the text. The exact relationship between the two types, which occur together, is not known. Lower arrows indicate mineral and gas exchange.

waved bright red fans of feathery tentacles, sweeping the water for oxygen and particles of food. Their tubes ranged up to two or three meters in length, while the worms living inside them reached about half that and were four to five centimeters thick. These strange, primitive creatures had no eyes and no mouth—in fact, no gut at all. Apparently, they absorb food directly from the seawater into their bloodstreams, where bacteria help them to digest it. The tubeworms were found only in the relatively warm waters close to the vent openings. Ringing the central forest of tubeworms, and often mingled in among them, large beds of clams and mussels grew in the chinks and folds of the rocks. The clams reached amazing sizes—up to one foot long. Though their shells were a pale, chalky white, their bodies were an unusual deep red color, due to the presence of high levels of hemoglobin in their blood.

Scattered in among the beds of clams and mussels were thriving clusters of whelks, barnacles, leeches, and limpets. At the fringes of the oasis, there were sea anemones and odd little creatures: small, 5-centimeter spheres made up of hundreds of soft members; they resembled dandelions just gone to seed, and so they were called. Later, it was found that the "dandelions" were tiny siphonophores, jellyfish-like relatives of the Portuguese Man-of-War. Occasionally, anemones were found down inside the vent opening itself, amid the warm-water flow.

Ranging back and forth over the whole of the community, scuttling over the rocks in the darkness, were numerous species of crabs and other crustaceans. Many of them were blind, their eyes having atrophied and vanished in the course of adapting to their strange environment.

The investigators doubted that simple warmth could account for the lush gatherings about the vents; the anemones were evidence of that. They existed both inside the vents and beyond the warmth of their waters, doing equally well in either place. In fact, life has been found to exist all across the ocean beds, despite the pressure, the darkness, and the cold, although it is usually sparse and widely scattered. Conditions on the sea floor do not prevent life; they only make it more difficult. The real limitation for organisms at those depths is the food supply—a source of energy.

Usually, deep-sea creatures depend on the slow rain of detritus from the more hospitable surface regions; ultimately, they, too, are dependent upon photosynthesis. The benthic communities surrounding the geothermal vents were oddly out of place; the density and diversity of life in them was astonishing. They should not have been able to thrive so on the limited resources available. What was the source of food, of energy, that allowed them to flourish?

From the first, the researchers suspected the presence of chemosynthetic (as opposed to photosynthetic) bacteria; such organisms would relish the warm, mineral-laden geothermal waters. The milky-blue color of the vent outflows provided a clue: it suggested that such bacteria were present, oxidizing the hydrogen sulfide dissolved in the water into sulfate and elemental sulfur. Chemosynthetic, sulfur-metabolizing microorganisms have been found in many places; these bacteria utilize the energy extracted from the oxidation process to fix carbon dioxide in a manner analogous to the way green plants utilize sunlight. Other substances that these and related species of bacteria also devour include hydrogen, ammonia, nitrates, iron and manganese ions, as well as sulfate and sulfur itself, via different metabolic pathways.

Laboratory studies of samples taken from the vents soon confirmed these early suspicions; in fact, over *two hundred* strains of sulfur-oxidizing bacteria have been identified, exhibiting a wide range of metabolic types. Most of them preferred a low concentration of oxygen, and some anaerobic forms (those that cannot tolerate the presence of oxygen) also were found, extracting their energy from the sulfur compounds by reduction rather than oxidation. Many of them grew well in laboratory cultures without added sources of nitrogen, although, strangely enough, few of them exhibited the ability to fix this element on their own.

It quickly became apparent that this large, complex bacterial community also serves as an ideal source of food and energy for the benthic organisms outside the vents. The bacteria fix carbon dioxide and other nutrients into organic compounds, acting as the primary producers in establishing a unique food chain totally independent of the sun-based, photosynthetic ecology with which we are familiar. The ultimate source of energy here is geothermal: hot rocks and water.

As the hot, geothermal waters begin rising up through the cracks and crevices in the newly formed rock, they take on their load of dissolved minerals and gases. The ever-present bacteria, until now quiescent or dormant inside spores, re-awaken and begin to multiply rapidly, oxidizing the hydrogen sulfide and other mineral nutrients. As the new vents begin to flow, the bacteria grow in thick mats in the chinks and crevices inside, but some of them are ejected; sometimes a portion of the mat peels away in the warm current and it, too, is ejected from the vent. The thriving bacteria cloud the waters leaving the vents, providing a rich suspension of food for the organisms that have begun to cluster around the openings. The concentration of organic matter in the water close to the vents is 300 to 500 times that in the surrounding depths, so cold and dark, and up to four times that of even the sun-drenched surface levels.

Filter-feeding organisms like the tubeworms and barnacles begin to thrive on the rich bacterial harvest, as do the clams and mussels feeding on the high-density soup of microorganisms. The clams grow at rates of up to four centimeters per year, 500 times faster than their smaller cousins elsewhere in the depths, away from the vents.

Secondary consumers, attracted to the growing bonanza, begin to arrive. Crabs and other crustaceans feed on the carcasses of the filter-feeders, or on the drifting clumps of bacteria torn loose from the dense mat inside the vent. Investigators also have found one small shrimp-like crustacean which, in place of eyes, had tiny comb-like structures atop its eyestalks, which are probably used as rakes to harvest growths of microorganisms from the rock surfaces. Even free-swimming fish are attracted, feeding head down in the vent outflow on the emerging clumps of bacteria. Predators are few, but creatures such as the octopus have been seen at the sites.

While the rocks below are hot, and the circulating waters flow, life is easy and food abundant. The larvae of the vent dwellers drift away on the slow undersea currents; some settle nearby, others drift longer in search of fresh, newly-opened vents. But the rocks below begin to cool, or the circulation of the ocean waters is cut off, and eventually the vent flows decrease, sputter, and die out. In a few years, all that is left to mark a once-thriving oasis is a field of empty, bone-white clamshells, dissolving slowly in the cold, dark waters. Investigators have found many of these abandoned "ghost vents."

The hydrothermal vent communities and their unique ecology caused quite a stir in the biological sciences, and the discovery received widespread public attention. Thoughtful workers in the field noted that many of the sulfur-metabolizing bacteria were *archaebacteria*, a class of microorganisms whose roots go back very nearly to the origin of life on Earth, predating green plants, the accumulation of atmospheric oxygen, and the rise of multicellular organisms. Was it possible that these ancient bacteria were utilizing geothermal energy sources even before photosynthetic forms evolved?

Other interest centered on the higher organisms of the vent communities, many of which were entirely new to science, unique to their specialized environmental niche. One type of limpet found at the vent sites, for example, was found to be an ancient filter-feeding type previously known only through fossils. The huge clams and tubeworms, though related to known species, were quite different and vastly larger than anything previously discovered. It was obvious that these creatures were well adapted to their unusual environment. The numerous crabs that swarmed over the vent oases, for instance, required a crushing pressure of *at least* 125 atmospheres just to survive; they did not last long without it. The water pressure at the depths of the vents themselves was over twice that, at 265 atmospheres. (Sea-level atmospheric pressure is, by definition, one atmosphere.)

Just how widespread were such communities on the ocean floor? The question was a puzzling one, and it has not yet been answered. No sign of such benthic oases had been seen in the Atlantic or the Caribbean, but the area explored there had been very small. Thus, it was with some excitement and anticipation when, during the first half of 1979, a new series of dives—which included researchers from not only Oregon State and Woods Hole, but the Scripps Institution of Oceanography, the Office of Naval Research, and a host of others from the United States, France, and Mexico—began on the East Pacific Rise just off the coast of Mexico, at the mouth of the Gulf of California. This time, a team of marine biologists was included on the expedition rosters, and once again what the divers found was fascinating and unexpected.

The investigators were delighted when they discovered a new field of the strange hydrothermal vent communities, very similar to but subtly different from those explored at the Galapagos Spreading Center. But the differences were minor, easily explained by geographic variation. Still, interest was high. Extending their explorations across the surrounding sea floor led to the discovery of an entirely new type of vent. Unlike the gentle, warm-water flows previously encountered, these new vents gushed from the congealed lava formations with scalding violence. Streams of superheated water, flowing at rates of several meters per second, jetted from openings atop tall mounds of mineral deposits, colorfully mottled and bizarre (Figure 1-B.).

The waters of these undersea geysers have been measured at temperatures of up to 360°C, well above the normal boiling point of H_2O. It is the fantastic pressure at those depths that prevents it from flashing into stream; in fact, it has been calculated that as seawater circulates down into the rocks and cooling magma far below, where the pressure is even greater, the geothermal waters are heated to a searing 450°C. Yet they remain liquid, and begin rising back up through the faults and seams in the rock, exchanging minerals and gases.

The superheated seawater thus carries up a tremendous load of dissolved minerals including copper, iron, zinc, cobalt, lead, silver, cadmium, and others, as well as sulfur compounds and the various trace gases. As the scalding stream jets from the vent, it encounters the frigid ambient seawater, at 2°C; it mixes rapidly and cools within a foot or two of the opening. The dissolved minerals begin to precipitate out, settling to the sea floor near the vent and building up a tall mound of deposits. The mounds often show a strange mottling of colors—the yellows, ochres, and browns of the sulfates of iron, copper and zinc. The precipitates form layered crusts; samples of the dull copper and zinc sulfide shells often break open to reveal glittering crystals of fool's gold—chalcopyrite or iron pyrite. The geysers are dubbed black or white "smokers" because of the murky clouds of mineral-laden water that surround them.

The interiors of the mounds are often honeycombed with the buried shells of tubeworms and clams, for these "smokers" support rich communities of life very similar to those surrounding low-velocity, low-temperature vents nearby and at the Galapagos site. Numerous samples were taken, and laboratory studies have confirmed that there are thriving bacterial communities surrounding these high-temperature vents as well.

But investigators such as John Baross, Marvin Lilley, and Louis Gordon at Oregon State University were severely puzzled when they began to detect viable microorganisms in samples taken directly from the scalding 350°C jets of water.

At first, they suspected their samples had been contaminated by the surrounding seawater; obtaining pure samples with the remotely controlled manipulators of the *Alvin* was not easy. Still, they were aware of those unusual sulfur bacteria that exist in boiling hot springs elsewhere, such as those which color the steaming pools of Yellowstone Park; and so they ran a check.

The incubation was carried out under similar conditions: pressure was kept at one atmosphere, and the water bubbled at 100°C. The bacteria in the culture went happily about their business, feeding and multiplying upon a medium containing manganese, ferric, and sulfate ions as energy sources, and giving off measurable quantities of methane, carbon monoxide, hydrogen, and traces of nitrous oxide. The growing bacterial community included both oxidative and anaerobic species, and it was found that not only did they tolerate the boiling heat, they actually *required* it: the bacteria would not grow at temperatures below 70° to 75°C.

Yet the investigators were skeptical that the bacteria had actually come from *inside* the smoker vent; the temperature was much too high. But to be absolutely sure, Baross and a co-worker, Jody Deming, ran a second series of incubation tests under conditions close to those of the site where the samples had been collected: a crushing 265 atmospheres of pressure and a temperature of 250°C.

Incredibly, the bacteria continued to thrive, even multiply. Nothing like it has ever been observed before. As of the autumn of 1983, two morphologically distinct organisms have been isolated which not only tolerate such extreme conditions, but carry on a wide range of physiological activities, apparently unperturbed, utilizing the oxidized and reduced metals, sulfur compounds, and dissolved gases in the searing geothermal waters to carry out their metabolic activity. The implications of this discovery are intricate, fascinating, and far-reaching, touching many fields.

First of all, we must once again extend the range of conditions under which we *know* life can exist, and that is always something of great interest. A few years ago, it was generally thought that organic life could not exist outside the rather narrow range between 0° and 100°C in which water remains liquid under the conditions prevailing on the surface of the Earth. Since then, we've found organisms living in the highly saline brine ponds of Antarctica, surviving at temperatures well below the normal freezing point of water. Now, we've found other forms that thrive happily at the other extreme, living in hot water at 350°C or more. What sort of proteins and enzymes can resist being denatured at that temperature and still function? How do these microorganisms prevent the heat from disrupting the delicate machinery of their metabolism altogether?

Obviously, it's time to change our thinking. Perhaps a more useful generalization would be to assume that we'll find life wherever there's a source of energy and liquid water, regardless of the temperature.

This is an interesting notion, and it leads us toward some intriguing speculation. Given the presence of living bacteria in these deep-sea, high-temperature hydrothermal vents, one wonders if perhaps similar bacteria (and even higher organisms?) may have penetrated deep into the searing interior of the Earth's crust, at least as far as water can also penetrate and remain liquid. Life in such an environment would be strange indeed.

Workers in the field do not discount the possibility. Oceanographers and geologists have begun to rethink their theories on the origins of those dissolved minerals and gases flowing from the vents. How much of this is due to simple geochemistry, and how much is produced by the activity of microorganisms inside the rocks? It is possible that a great deal of the methane, carbon monoxide, and hydrogen is due, not to water-rock reactions, but to biological activity *beneath* the planet's crust. This is another example of the intricate, interlocking relations between the Earth's lithosphere, atmosphere, and biosphere which shape the environmental conditions on our world.

The discovery of these thermophilic (heat-loving) bacteria and the associated oases of benthic organisms has also raised some interesting new questions concerning the origin and early evolution of life on this world, and perhaps on others. The theories in general circulation today are based on the idea that pre-biological organic molecules were created from the gases of the atmosphere, via mechanisms driven either directly or indirectly by the sun. This is still the accepted scenario, but now the possibility exists that early forms of life on this planet may have developed independently of such mechanisms, or alongside them. It is probable that both pathways played a role in the origin and future course of evolution of terrestrial life, but one wonders: just how long have those ancient archaebacteria been thriving in these hidden geothermal niches? How long has the Earth had an ecology existing independently of photosynthesis?

So far, the only fossil clues we have were discovered in Oman last year. Fossil remains of worm tubes similar to those of the worms seen at the vents today have turned up embedded in massive sulfide deposits from an Omani mine, and are at least 95 million years old.

Finally, the implications of these findings for the possibility of life existing on other worlds are really quite exciting. We might not even need to look beyond our own solar system. While Venus is still thought to be too hot and too dry for any life to exist there, we could now reasonably speculate that perhaps some hardy sulfur-metabolizing bug might be tailored to the harsh conditions on that planet, if not on the surface, then perhaps seeded high up in the dense carbon dioxide atmosphere, where thin clouds of liquid water and sulfuric acid smog are found.

To take our speculation a bit farther afield, let's consider Europa, the second closest Galilean satellite of Jupiter. It's thought that the interior of Europa may be heated by a milder version of the same tidal resonance "squeezing" that has melted the interior of Io, driving the incredible volcanic activity on that minor world. Now, Europa differs from Io in that its composition includes a large component of ices, chiefly water ice, where Io seems to have vented most of its volatiles into space. Europa might be thought of as a giant rocky snowball; a considerable fraction of its radius is thought to be ice, overlaying a rocky core.

But is all the water frozen? There is some evidence, still sketchy and tentative as yet, that the same gravitational tug-of-war that causes Io's volcanic plumes may also have heated the interior of Europa, and melted some of the ice overlaying the rocky core. By this reasoning, the Voyager images of Europa show only a relatively thin (1–10 km) crust of ice, beneath which lies a mantle of liquid water perhaps 100 kilometers deep, the largest ocean in the solar system beyond the Earth. The theory goes a long way toward explaining the absence of impact features on Europa, and its fractured appearance. The fissures that cover the face of Europa could be cracks in its icy crust, where occasional "volcanoes" of liquid water erupt, covering over any impact craters and scars (Figure 2).

Others have gone so far as to speculate that some primitive form of life may have evolved on Europa while Jupiter was still warm from the heat of its formation, and might have developed a sort of photosynthesis, utilizing the weak sunlight that penetrates the thin ice at newly opened fissures. But need such life cluster beneath

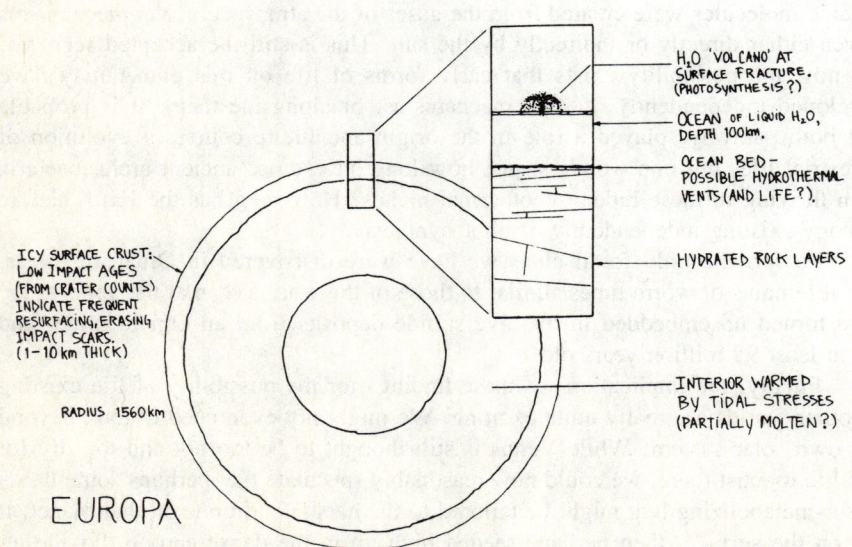

FIGURE 2. One possible configuration of the interior of Europa. The interior of neighboring Io is thought to be completely molten due to tidal resonance heating. Europa may experience a similar effect of a lesser magnitude.

these occasional openings, existing on the near invisible rays of the distant sun? As we've seen, photosynthesis isn't the only way to go.

If Europa does have an active interior beneath a mantle of liquid water, it is not beyond our imagining that vents similar to the ones we've found in our own seas stir the cold waters of a deep, sunless ocean on Europa with warm geothermal flows. To go one step further, we might imagine some rather simple ecologies springing up, just as we find them huddled around hydrothermal vents here on Earth.

Speculation? Perhaps. We really ought to go and take a look. But it's certainly possible that we'll find life in such strange environments, born of some combination of sunlight, hot rocks, and liquid water.

Part Three

WHAT IS THIS THING CALLED MAN?

H. Keith Henson

MEMETICS AND THE MODULAR MIND

AUGUST 1987

IN his *Foundation* stories, Isaac Asimov proposed a future science called "psychohistory," in which the collective behavior of human populations could be predicted with high precision. In our time, the social sciences are often viewed as sharply different from the physical sciences because they *cannot* do much predicting. Is this an inherent limitation on the social sciences, or might it be possible to put them on a truly predictive basis by means that have not been formulated yet? There are a number of lines of research suggesting that it might. One of them is based on the "meme": a concept created by analogy with the gene and describing an entity supposed to behave in a somewhat similar way.

H. Keith Henson was one of the founders, and the first president, of the L5 Society, which has since become part of the National Space Society. He describes himself as a carrier for several highly infectious memes relating to space colonies, nanotechnology, personal computers, and cult-watching. ■

SCIENCE fiction writers do not always manage to stay ahead of science. One significant concept showed up in the scientific literature 13 years before Charles

Sheffield and Arthur Clarke simultaneously wrote stories that incorporated the "Skyhook" or "Beanstalk." But in projecting a science of social prediction, SF writers have been far ahead of the scientists. Isaac Asimov based the entire Foundation series on "Psychohistory." Robert Heinlein developed the theme of predicting social movement in his Future History stories, especially in *Revolt in 2100, Methuselah's Children*, and in the unwritten saga of Reverend Nehemiah Scudder.*

Science fiction aside, we don't have a science of social prediction. Until recently, we haven't even had much in the way of theories. Our continual surprise at the development of cults, religions, wars, fads, and other social movements is a notable exception to the steady progress humans have made in building better models of our environment. When you consider the suffering associated with some social movements, our lack of good models must be considered a major deficiency.

A successful theory of the development of social movements will have to provide a unifying theory for events that make up much of the evening news. It will have to discover common features that lie behind the diverse trends causing problems in Nicaragua, South Africa, Northern Ireland, and the Middle East. A good theory should be able to evaluate the danger or lack of danger from the LaRouche organization, whose accidental win in the Democratic primary forced Adlai Stevenson III to run as an independent in the Illinois governor's race. (This cult more recently made the news when the FBI raided its offices in the wake of alleged massive credit card frauds.) It should be able to produce a plausible model for the breakup of the Rajneesh cult (whose Bhagwan Shree Rajneesh accumulated 93 Rolls Royces before abandoning his Oregon community). The theory should be able to predict the conditions under which Turkey will be subverted by a fundamentalist version of Islam similar to that which led to so much grief in Iran.

A tall order! But an emerging field of study, *memetics*, holds just such promise. Sometimes thought of as "germ theory applied to ideas," memetics is an outgrowth of evolutionary biology. It provides models where social movements are seen as side effects of infectious ideas that spread among people in a way mathematically identical to the way epidemic disease spreads. It has been noticed, for example, that use rates for various drugs, most recently "crack," have closely followed epidemic-like curves that seem to be as oblivious to the efforts of authorities as the Black Death was in 1348. At a deeper level, research in neuroscience and artificial intelligence is starting to develop an understanding of why we are susceptible to "infectious information," both the benign and the deadly.

As useful as these models may be, they are not without the potential to seriously affect our cherished institutions. A good understanding of the mechanisms

---

* "First Prophet," President of the United States, destroyer of its Constitution, and founder of the Theocracy. If this makes you vaguely uncomfortable, it is probably because you have been reading about fundamentalist preacher/presidential candidate Pat Robertson. As the Ayatollah Khomeini recently demonstrated, fundamentalist religion and politics can make a nasty mix.

of our minds and the dynamics that underlie the spread and persistence of *any* social or political movement has the potential to forever alter the way we think about all other social movements, including those of our own culture, religions, and nation. When viewed from the perspective of tolerance that has been developing in Western culture since the Renaissance, the changes in outlook seem to be positive, but it would not surprise me to find memetics condemned from the pulpit even more than evolution has been.

Memetics comes from "meme" (which rhymes with "cream"), a word coined in purposeful analogy to gene by Richard Dawkins in his 1976 book, *The Selfish Gene*. To understand memes, you must have a good understanding of the modern concepts of evolution, and this is a good source. In its last chapter, memes were defined as replicating information patterns that use minds to get themselves copied much as a virus uses cells to get itself copied. (Dawkins credits several others for developing the concepts, especially the anthropologist F. T. Cloak.) Like genes, memes are pure information.* They must be perceived indirectly, most often by their effect on behavior or by material objects that result from behavior. Humans are not the only creatures that pass memes about. Bird songs that are learned (and subject to variation) and the songs of whales are also replicating information patterns that fit the model of a meme. So is the "termite-ing" behavior that chimps pass from generation to generation.

"Meme" is similar to "idea," but not all ideas are memes. A passing idea which you do not communicate to others, or one which fails to take root in others, falls short of being a meme. The important part of the "meme about memes" is that memes are subject to adaptive evolutionary forces very similar to those that select for genes. That is, their variation is subject to selection in the environment provided by human minds, communication channels, and the vast collection of cooperating and competing memes that make up human culture. The analogy is remarkably close. For example, genes in cold viruses that cause sneezes by irritating noses spread themselves by this route to new hosts and become more common in the gene pool of a cold virus. Memes cause those they have successfully infected to spread the meme by both direct methods (proselytizing) and indirect methods (such as writing). Such memes become more common in the culture pool.

The entire topic would be academic except that there are two levels of evolution (genes and memes) involved and the memetic level is only loosely coupled to the genetic. Memes which override genetic survival, such as those which induce young Lebanese Shiites to blow themselves "into the next world" from the front seat of a truck loaded with high explosives, or induce untrained Iranians to volunteer to charge Iraqi machine guns, or the WW II Kamikaze "social movement" in Japan, are all too well known. I have proposed the term "memeoid" for people

---

*The essence of a gene is in its information. It is still a gene "for hemoglobin" or "for waltzing behavior in mice" whether the sequence is coded in DNA, printed on paper, or written on magnetic tape.

whose behavior is so strongly influenced by a replicating information pattern (meme) that their survival becomes inconsequential in their own minds.

For a vivid example we can hark back a few years ago to Rev. Jim Jones and the People's Temple incident, where 912 people, including Jones, died of complications—poison and gunshot wounds—induced by an information disease. The Children's Crusades of the middle ages were larger and more lethal; only 2 of 20,000 returned from one. The mass suicide in the first century by the Jews at Masada is a clear example of information patterns in people's minds having more influence over their behavior than the fear of death.

A more seductive example of a social movement set off by a lethal meme comes from South Africa. In the 1850s, a meme (originally derived from a dream) led to a great sacrifice by the Xhoas people during which they killed their cattle, burned their grain, and refrained from planting in the belief that doing so would cause their ancestors to come back from the dead and expel the whites. At least 20,000 and perhaps as many as 60,000 starved when the predicted millennia of plenty failed to arrive. Known as the Cattle Killing, it was not a unique response for a primitive society being displaced by a more technically advanced one. The "Ghost Dancers" phenomenon among American Indians was a similar response.

Memes that bring about suicidal behavior are at least self-limiting. Those which induce one group of people to kill another are much worse, and the social movements they induce are often much larger. The scope of the social movement known as the Inquisition is seldom mentioned in history textbooks, but:

> The number of victims claimed by the witch-hunts, which lasted for three hundred years, is reckoned by historians to be between five and six million people; it therefore caused more deaths than all the wars waged over the period. . . .
>
> It is only when one takes into account the brutal, pitiless, expression of mass-mania, and that a belief in the devil, his traffic with witches and warlocks, was constantly being fanned anew by the Church . . . that it is possible to gain any measure of understanding. . . .*

The depredations and brutality of the Inquisition were about typical of deadly memes stemming from religions or closely related social movements such as Marxist-Leninist communism.

In the last decade, the people of Kampuchea were infected with an antiintellectual, agrarian utopian meme clearly mutated (in the minds of Pol Pot and his close associates) from the Vietnamese variation of the communist meme. They were Eric Hoffer's "True Believers" of the most extreme stripe. The resulting social movement was a massive self-genocide. Over one third of the population of

---

Five Thousand Years of Medicine by Gerhard Venzmer, Tr. Marion Koenig, Taplinger Publishing Co., NY 1968 pg. 163.

Kampuchea, including almost all of the city dwellers and the educated, died before the Vietnamese (embarrassed by news stories of rivers clogged with bodies) invaded and put a stop to the killing. Many more would have died had the social movement run its course without interference. Kampuchea will take decades to recover, but "'tis an ill wind . . .'' The people of Thailand, with a front seat on the slaughter, seem to have lost all sympathy for their own related social movements.

History classes have made us more aware of the genocidal depredations resulting from the "master race" meme that was part of the Nazi meme complex. Considered from the viewpoint of memes, Hitler was less a prime mover than a willing victim of this particularly nasty and pervasive variety of information disease. Had plague struck Germany in the '30s instead of Nazism, we would have understood it in terms of susceptibility, vectors, and disease organisms. What did happen may soon be modeled and understood in terms of the social and economic disruptions of the time increasing the number of people susceptible to fanatical beliefs, just as poor diet is known to increase the number of those susceptible to tuberculosis. For vectors, we have personal contact, the written word, radio, and amplified voices substituting for rats, lice, mosquitoes, and coughed-out droplets. A pool of "submemes," many of them ancient myth, contributed to the syncretic Nazi meme in much the same way mobile genes contribute to the virulence of the influenza viruses.

Nazism was not the only fanatical movement growing and evolving in the fertile social media of Germany between the wars. The Marxist-Leninist meme was a visible competitor in the early period. Even though most of those infected with the Nazi meme were conquered or killed and Nazism became a suppressed meme, it cannot be said to have died. As a replicating information pattern that has gone through a great deal of evolutionary honing, it is still successful in infecting a few susceptible people today.

A fascinating footnote to the German experience with Nazism and its horrors happened in 1969 when Ron Jones, a teacher in Palo Alto, exposed a high school history class to an intensive, five-day experience with the ideas that made up the Nazi meme. The experience of that week was originally published as "Take as Directed" in *The CoEvolution Quarterly*, Spring '76 and a few years ago was made into a TV movie, *The Wave*. Over four days, Jones introduced and drilled his students in concepts of Strength Through Discipline, Community, Action, and Pride. (The fifth day was devoted to showing them how easily they had started to slip into the abyss.) The enthusiasm with which most of the class adopted the memes and spread them to their friends, swelling a 40-student class to 200 in 5 days, made it one of the most frightening events the teacher had ever experienced. Given the track record of the Nazi meme, the mini-social movement his experiment set off is no more surprising in retrospect than the medical effects would have been if the teacher had sprayed smallpox virus on the class.

An empirical characteristic of large, long-lived religious movements or related social movements (at least in the West) is a scripture or body of written material. This may function to standardize the meme involved or at least slow its evolution as the number of people infected with it grows. From Scientology right back to the

Hindu Vedas, I can think of no counter-examples. Social movements involving more than a few thousand people or lasting more than a few years may have been rare before writing came along.

It is possible that the breakup of the Rajneesh cult was related to its lack of an organized written scripture at a critical juncture. The memes that were the origin of that particular social movement were characterized by considerable instability; that is, parasitic memes arose out of the local culture soup at short intervals. Some of them (tapping phones) made a kind of paranoid sense, but poisoning salad bars at restaurants with Salmonella bacteria in the hope of influencing local elections made no sense. The group seems to have amplified individual crazy impulses at the expense of propagating the meme.

I have noticed several features of the social movements that derived from really dangerous memes. One is self-isolation of the infected group or at least new recruits from the rest of society. This need not be an "intelligent" action taken by the "leaders." There may be no more thought involved than the selection of dark moths in industrial England. The "fanatic cult" memes which incorporate isolation are the ones we observe; those which do not incorporate isolation are like light moths, gone and not observable.

In the case of the Soviet Union, the cult-like communist meme survives in a society largely isolated from the rest of the world. In recent years the isolation may have resulted from reasoned considerations about the fragility of the communist meme in open competition with other memes. A more parsimonious view would note that without originally having a strong isolation component, the communist meme would have had no more social influence in the USSR than it has had in, say, France.*

Isolation makes possible exposure to a single meme (or meme set) many times a day for months or years without much contact with other memes. Exclusive exposure to one meme (also known as brainwashing) induces a "dependent mental state" in some people.

Thankfully, most of us have not experienced the dependent mental state firsthand, but we have all seen such people on the news programs boarding buses for the front in Iran, or been harassed by them in airports, or had them knock on our doors and try to infect us. It is clear that the people who suffer from extreme cases of "information disease" have lost much of their ability to take care of themselves or their children. Truly dedicated people often fail to replace themselves, since too much of their life energies are channeled into propagating the infecting meme. One example comes from the largest subdivision of Christianity, where celibacy for its most dedicated has long been institutionalized. The Rajneesh cult practiced the opposite of celibacy but discouraged births to the point of sterilizing the barely pubescent female children of its resident members.

Given that memes have been interfering with our reproduction for a long

---

*The ferment in the USSR today is certainly consistent with this point.

time, one must wonder why humans are still so susceptible to information diseases. The answers to such questions are starting to come from research in artificial intelligence (AI), neuroscience, and archeology. It is becoming apparent that our vulnerabilities are a direct consequence of the way our minds are organized, and that organization is a direct consequence of our evolutionary history.

Marvin Minsky (a principal founder of AI) and Michael Gazzaniga (one of the major workers in split-brain research) have independently come to a virtually identical model of the mind. Both view minds as vast collections of interacting, largely parallel (co-conscious) modules, or "agents." The lowest level of such a society of agents consists of a small number of nerve cells that innervate a section of muscle. A few of the higher-level modules have been isolated in clever experiments by Gazzaniga, some of them on split-brain patients.

One surprise from this work is that we seem to have our mental modules arranged in a way that guarantees we will form beliefs. *What* we believe in depends, at least in part, on what we are exposed to and the order in which we are exposed. Gazzaniga argues that we slowly evolved the ability to form beliefs because the ability provides a major advantage in surviving. Being able to infer, that is to form new beliefs, and to learn, in the sense of acquiring such beliefs from others, was a major advance over learning by trial and error. Being able to pass the rare new ways our ancestors found for chipping rock or making pots from person to person and generation to generation was vital in allowing humans to spread over the earth.

But as this ability became the norm, communicating human minds formed a new "primal soup" in which a new kind of non-biological evolution, that of replicating information patterns or memes, could get started. A wide variety of competing memes has evolved in the intervening seventy thousand years or so. It should not be surprising that the survivors of this process, like astrology or religions, are so effective at inducing their hosts to spread and defend them. It is also plausible that in the tens of millennia since memetic evolution became a major factor, there has been a biological co-evolution. The parts of our brains that hold our belief systems have probably undergone biological adaptation to be better at detecting dangerous memes and more skeptical about memes that result in death or seriously interfere with reproductive success.

This type of co-evolution is known as an "arms race" to biologists. One such biological arms race has resulted in almost perfect egg mimicry by the cuckoo and in correspondingly sharp visual discrimination in the birds it parasitizes. By analogy, while we get better at spotting dangerous memes, the memes may be evolving to be more effective at infecting us. Advancing technology (which itself is an improving collection of memes) changes the environmental conditions where memes survive or fail as well. The modern telephone system and the tape cassette player were major factors in the takeover of Iran. It has been argued that the rise of the Nazis depended strongly on radio reaching a previously unexposed and unsophisticated population. Exposure to modern advertising may be one factor which makes a television broadcast by Lyndon LaRouche attacking (among others) the L5 Society so absurd that tapes of it are used as entertainment at L5 parties. He might have been taken seriously in the '30s.

I have picked dangerous examples for vivid illustrations and to point out that memes have a life of their own. The ones that kill their hosts make this hard to ignore. *However, most memes, like most microorganisms, are either helpful or at least harmless.* Some may even provide a certain amount of defense from the very harmful ones. It is the natural progression of parasites to become symbiotes, and the first symbiotic behavior that emerges in a proto-symbiote is for it to start protecting its host from other parasites. I have come to appreciate the common religions in this light. Even if they were harmful when they started, the ones that survive over generations evolve and do not cause too much damage to their hosts. Calvin (who had dozens of people executed over theological disputes) would hardly recognize Presbyterians three hundred years later. Contrariwise, the Shaker meme is now confined to books, and the Shakers are gone. It is clearly safer to believe in a well-aged religion than to be susceptible to a potentially fatal cult.

History doesn't change, but our interpretation of it can. For example, the contemporary "causes" of historical epidemics (such as the miasma theory) have been totally supplanted by germ theory explanations. Before germ theory came along, memes of causality for epidemics were remarkably stable. The "explanation" for the Black Death of 1348 was still in use for the Philadelphia Yellow Fever epidemic of 1796. Similarly, various "explanations" for wars have been with us for hundreds of years.

Memetics provides an interesting alternate way to analyze recent wars and the roots of current disputes. In this view, the ultimate (though unaware) protagonists of World War II were memes such as the Nazi "master race" and the Marxist-Leninist meme (MLM). The current clash between the Soviets and the western world can be viewed as a meme conflict (for space in minds) between the religion-like, competition-intolerant mono-meme of communism and the western meta-meme of tolerance. While it is not a religion by any reasonable definition, the Marxist-Leninist meme is clearly in competition for the "belief space" in minds usually occupied by religious memes. It, and its more cultish offshoots, have the typical virtues and excesses of cult-stage religious memes. In an amusing twist, the "godless" communist meme is the more religion-like of the two in its battle for mind space with secular western culture!

Reviewers of an earlier draft of this article objected to my description of Soviet memes. Words like "tolerant" or "intolerant" have acquired a great deal of positive/negative connotation in the western world, but in describing memes, I am using them in the same way we would say that a mold colony is intolerant of a bacterial invasion. With respect to the belief system that dominates the meme pool of the other superpower, I am trying to be descriptive, not partisan.

If anything, I would think that understanding the memetic nature of religions and related movements like communism would defuse the emotional connections and substitute something closer to dispassionate understanding of the parasitic-to-symbiotic memes behind such social movements. It has had that effect on me. Many, even the most gruesome, features of communism are what they are simply because those features were (and are) necessary for the meme to exist in a world of competing memes. Isolation, for example, is a common feature of virtually all successful religious memes while they are in the cult stage. Anyone who has studied history knows that suppression of competitive memes by the power of the state is a common experience once a meme of this class has infected the leaders or they have been replaced by those infected.

And if the Christian religion was a mainstay of the aristocracy, serving to keep the peasants in place, Soviet Communism is no less supportive of its own hereditary elite. As a successful and persistent meme, that has appeal even to people who know the realities of its practice. It commands a certain grudging respect.

From a meme's viewpoint, tolerance of other memes is *not* a virtue; it is, in fact, a fatal characteristic for a particular meme, as memes inducing *in*tolerance to other memes would soon displace it. On the other hand, a meta-meme of limited toleration, even cooperation among memes is possible. The western metameme of tolerance seems to have emerged from an ecosystem of memes in much the same way that cooperative behavior has been modeled as emerging from an ecosystem of individuals.* In the area of meme tolerance the western world may be unique. We think of censorship as evil; where but in an advanced ecosystem of memes could such a strange idea have emerged?

I have recently had a lot of fun reading history to trace the development of the meta-meme of tolerance. This particular character of our ecosystem of memes has been developing at least since the writings of the Greeks and Romans were rediscovered during the Renaissance. Studying inactive pagan religions may have been the first step in developing tolerance for a variety of religious memes. The fragmentation of the dominant religion during the Reformation led to a series of largely indecisive religious wars in most of the major countries of Europe. Sheer exhaustion may have been one of the most significant factors in developing a grudging tolerance, which in these later times has taken on a patina of virtue in the division of our culture known as "liberal."

In this view, western culture is a vast ecosystem where memes of many classes engage in "fair" competition with each other. Attempts to subvert fair competition by changing laws or education (such as introducing "creation science" into schools) draw opposition from defenders of a wide variety of memes which have evolved within this environment. This model may provide testable explanations for both western culture's tolerance of intolerant memes (such as creation science and the MLM) and the hostility these memes evoke from various segments of the culture. David Brin's "Dogma of Otherness" in the April 1986 *Analog* prompted considering a memetic explanation for such peculiar ambiguities in our culture.

---

*See *The Evolution of Cooperation* by Robert Axelrod, 1984 Basic Books, NY.

Several current social movements are obvious candidates for examination with memetic theory. Given the available data, we may be able to predict the remaining course of the "non-literate graffiti epidemic," which has spread in the past 15 years from New York City to remote corners of the country. There are substantial financial reasons (such as the cost of mark-resistant walls) to want to know if scribbler behavior will be a limited epidemic or will become an endemic part of our culture.

Drug use, clearly a replicating pattern of behavior passed from person to person, is another "social movement" where the similarity to epidemic waxing and waning has been widely used by reporters, and noted without much explanation in a number of learned journals. If it were formally considered as an epidemic with memes as the infecting agents, the ways by which the behavior spreads might get more attention. Counter-drug programs might be evaluated in terms of how well they induce reasonable behavior. Some efforts in the past, especially those which wildly exaggerated the dangers of a drug such as marijuana, may have *increased* the behavior of taking other drugs. These efforts may have immunized those exposed against believing any official pronouncements about drugs.

Formal consideration of drug use as an epidemic of meme-induced behavior might also lead to the realization that the percentage of people susceptible to abusing most drugs is not all that large. (Cigarette smoking is an exception.) For example, most of the people I know who have tried cocaine don't care for it. Not liking the effect, they wouldn't use it if it were free. People who really like opiates aren't that common, either.

Part of my interest in memes stems from a ten-year (and continuing) experience of being infected with the space colony meme which developed in the minds of Gerard O'Neill and his students in the late '60s. (See "Memes, L5 and the Religion of the Space Colonies," September, 1985 and "More on Memes," June 1986, both in *L5 News*.) Memetics provides candidate explanations for why the space colony meme spread in the first place, why it is not making much progress now, and some insight into what might be done to revitalize the meme and actually accomplish the implicit goals.

From a recent survey of L5 members, there seems to be two main factors and a minor one that contributed to the attractiveness of the space colony meme. First was the "new lands" factor. We are the genetic and memetic heirs of people who moved into vacant areas of the planet. It should be no surprise that the prospect of new lands is an irresistible attraction to many people. This may also explain the higher than proportional membership in L5 from California, where the last of the restless pioneers piled up. The minor factor (suggested by Dale Skran of Bell Labs) is fear of random events such as nuclear war, asteroid impact, worldwide epidemics, or crazy social movements that could badly damage civilization or even extinguish human life. As Heinlein put it, one planet is too fragile a basket to put all the human eggs in. The other main factor was possibility of personal involvement, of going into space. Surprisingly this is still a very important factor. Some 60 percent of respondents to a survey at the 1986 L5 annual conference said they expected to live in space.

If the attractiveness of the space colony meme is in the prospect of large numbers of people being able to live in space within a short time, these factors are quite at variance with today's reality; since the Solar Power Satellite project bit the dust, there haven't even been any widely accepted proposals that would get us out there in the next 50–100 years. Since the space colony meme never had a fixed deadline, the lack of correspondence between the meme and reality hasn't hit as hard as "the day after" hits a millennial religion, but informal surveys of former L5 members indicate that lack of a timetable was an important factor in their becoming inactive. If we want to get out there, we need to tap a very large source of social energy. The biggest single source of social energy on the planet is the meme conflict between the MLM and the western metameme. There are ways this might be tapped to get us into space, but that would take another article.

The memes which embody the germ theory of disease emerged when they did partly because "the time was right." The work of von Leeuwenheok, Semmelweis, Spallanzani, and their less remembered colleagues established in scientific culture the background memes about microorganisms. Without these cooperating memes, the ideas of Pasteur and Koch could not have replicated. The tragic history of Semmelweis and his statistical work on childbed fever stands as an example of the failure of a meme to take root in a culture before the conditions are right for its spread, no matter how true or useful to humans it may be.

If most conflict in the world is an indirect effect of memes, memetics holds as much potential for reducing human misery as the germ theory of disease. Just being able to model the interaction between the Soviets and the West in terms of memes might go a long way toward making the world a safer place. It took at least 60 years for the germ theory of disease to be widely accepted, though, as anyone who has traveled much knows, it still has a ways to go in many parts of the world. What are the prospects in the near future for a similar acceptance of the meme-about-memes? If it were widely accepted, what changes could we expect to see analogous to public health? Would widespread awareness of infectious information make us less susceptible to dangerous memes? Can we separate ourselves from the memes that possess us?

Further exploration of the analogy between replicating information patterns and the ecosystems-epidemic models biologists have painstakingly developed for other purposes may provide badly needed insight into the origin and courses of social movements and the nature of meme competition/cooperation. If memetics develops soon enough, it may provide help in evaluating proposed solutions to current international problems, predict the course of troublesome social movements, and suggest solutions for conflicts between social movements. If this article succeeds in infecting you with the meme-about-memes, perhaps it will help you be more responsible about the memes you spread and less likely to be infected by a meme that can harm you or those around you.

POSTSCRIPT—1989

Lyndon LaRouche has now been sent to jail for credit card fraud. Cults such as this one can almost be defined by the central meme gaining ascendency in the minds of the infected over all other considerations, moral and legal.

Computer viruses are an additional analogy to the more destructive memes. While memes infect *human* operating systems, computer viruses and worms infect *computer* operating systems.

Sadly, the meme-about-memes is not spreading as fast as I would like. Those interested in helping spread it can contact me at:

1685 Branham Lane, #252
San Jose, CA 95118

or through email at:

hkhenson@cup. portal. com

or

keith@toad. com

REFERENCES

In addition to Dawkins's '76 and '82 books, there are a number of books and articles directly discussing memes. One that reached a large number of readers was Douglas Hofstadter's Metamagical Themas column "On Viral Sentences and Self-Replicating Structures" in *Scientific American* (Jan. 1983) and reprinted in his recent book. There are numerous supporting sources, and a reliable source indicates that a journal of memetics may be offered soon.

Bohannan, Paul. "The Gene Pool and the Meme Pool," *Science 80*, November 1980, pp. 25, 28.

Cloak, F.T., Jr. "The Causal Logic of Natural Selection: A General Theory," *Oxford Surveys in Evolutionary Biology* 3, article 6, 1986. In press. Preprints available from F.T. Cloak, Jr., 1613 Fruit Avenue, NW, Albuquerque, NM 87104.

Dawkins, Richard. *The Extended Phenotype: The Gene as the Unit of Selection*. W.H. Freeman and Company, Oxford and San Francisco, 1982. See esp. Chapter 6, pp. 97–117.

Dawkins, Richard. *The Selfish Gene*. Oxford University Press, New York, 1976. See Chapter 11, first use of "meme."

Drexler, K. Eric. *Engines of Creation*. Anchor Press/Doubleday, Garden City, New York, 1986. See esp. pp. 35–38 and other references to "memes" in the index.

Henson, H. Keith.

"Memes, L5 and the Religion of the Space Colonies," *L-5 News*, September, 1985.

"More on Memes," *L-5 News*, June 1986.

"Memes, Mental Parasites, and the Evolution of Skepticism," unpublished monograph.

"Original Sin and Liberal Guilt," *Cryonics*, in press.

Hofstadter, Douglas R. *Metamagical Themas*. Basic Books, Inc., New York, 1985. See esp. Chapter 3, pp. 49–69.

Stefik, Mark. "The Next Knowledge Medium," *The AI Magazine*, Spring 1986, Vol. 7, #1.

Wilson, Edward O. and Charles J. Lumsden. *Genes, Mind and Culture: The Coevolutionary Process*. Harvard University Press, Cambridge, MA, 1981. See esp. Chapter 3 and earlier definitions of "culturgen" and "epigenesis."

The following works do not use the word "meme," but their contents help elucidate human behavior and cultural evolution.

Baker, Sherry. "A Plague Called Violence," *Omni*, Vol. 8, No. 11 (August 1986), pp 42ff.

Chase, Stuart. *The Tyranny of Words*. Harcourt, Brace and World; New York, 1938.

Cloak, F.T., Jr., et al. "The Adaptive Significance of Cultural Behavior: Comments and Reply," *Human Ecology*, Vol. 5, No. 1 (1977), pp. 49–50 (with references appended to the monograph).

Cloak, F.T., Jr., "Is a Cultural Ethology Possible?," *Human Ecology*, Vol. 3, No. 3 (1975), pp. 161–82.

Conway, Flo and Jim Siegelman. *Snapping*. Dell Publishing, New York, 1979.

Gazzaniga, Michael S. *The Social Brain: Discovering the Networks of the Mind*. Basic Books, Inc., New York, 1985.

Kelly, Kevin. "Information as a Communicable Disease," *CoEvolutional Quarterly*, Summer 1984.

Minsky, Marvin. *Society of Mind*. Simon and Schuster, New York, 1986.

Nisbett, Richard, and Lee Ross. *Human Inference: Strategies and Shortcomings of Social Judgment*. Prentice-Hall, Inc., Englewood Cliffs, New Jersey, 1980.

L. Sprague de Camp

MAN'S BIOLOGICAL FUTURE

NOVEMBER 1980

IS mankind still evolving—and if so, toward what? A look at our past and present suggests a view of our future that is both thought-provoking and occasionally disturbing. Why disturbing? Well, it seems we now have a fair amount of ability to exert some control over our future evolutionary course. What, if anything, should we do with that ability?

L. Sprague de Camp has long been fascinated by such questions—and practically all others. He graduated from the California Institute of Technology with a B.S. in aeronautical engineering (just in time for the Great Depression) and later earned a master's degree in engineering and economics from the Stevens Institute of Technology. He has also traveled widely and has continued to educate himself in a wide range of fields. In addition to nonfiction books and articles in many of those fields, he is particularly well known for his distinctive contributions to science fiction and fantasy. He is one of the very few writers ever honored with a Grand Master Nebula for lifetime achievement in science fiction. ■

IN a science fiction classic of the Golden Age—"Alas, All Thinking!" by Harry Bates, in *Astounding*, June 1935—the hero goes by time travel three million years into the future. He finds that mankind has dwindled to thirty-five super-geniuses whose skinny little bodies cannot hold up their huge heads without props, like the clamps that Victorian photographers used to keep their subjects from moving during a time exposure. All these futurians do is to sit and think. Horrified, the hero massacres the lot, which does seem a little bigoted of him.

Other writers have caused human beings to develop into supermen or telepaths, or, after an atomic war, into freaks with three eyes or two heads. From all we know of mutations, nothing so picturesque is probable. Then what is likely to happen to man as an evolving organism? A look at our past may give some hints about our future.

Up to the Agricultural Revolution of ten or twelve thousand years ago, man developed much as other life forms do. We have evolved during the preceding ten or twenty million years from small, bipedal, ground-living man-apes like those found fossil in Africa. We have since never become perfectly adapted to walking upright, as witness hernias, varicose veins, and fallen arches.

During the last hundred thousand years or so, our species, under the pressure of different environments, split into three main races: the black or Negroid, the white or Caucasoid, and the yellow or Mongoloid, together with a few small groups like the African bushmen and the Australian aborigines, which do not fit into any of the major races but are classified by anthropologists as separate minor races of their own.

The main racial differences are adaptations to different climates: the Negroid's dark skin, woolly hair, abundant sweat glands, and long-limbed shape to African heat and sun; the northern European's pale skin, full beard, and long, narrow nose to the cool, damp air and sunless skies of northern Europe; the Mongoloid's smooth, fat-padded face, coarse, straight hair, and stocky build to the bitter winters of Manchuria and Siberia. Different races also differ in their immunity to diseases, according to what they have been most exposed to. Thus Negroids resist malaria, on the average, better than Caucasoids; while in the case of tuberculosis it is the other way around.

For millions of years, our forebears lived in tiny, isolated bands of hunters and food gatherers. Three evolutionary forces molded them: mutation, selection, and genetic drift. Mutation, which I shall come back to, is, of course—as nearly all the present readers know—a sudden change in the mechanism of heredity.

Selection is the "survival of the fittest" of which Darwin's follower Herbert Spencer wrote. The fittest are simply those who have the qualities they need to leave more offspring than others of their kind. Fitness can take the form of being stronger, healthier, cleverer, brisker, more fertile, more aggressive, better at co-operation with others, or any combination of these qualities.

Some human traits that do not fit well into civilized life may be explained on the ground that they helped in our survival as hunters. An example is the built-in factiousness and quarrelsomeness that leads any large group to divide into factions on almost any pretext—racial, religious, national, linguistic, cultural, or even sporting—and fight it out. This trait may reflect the need for a hunting-gathering band, when it gets too large, to split amoeba-like, so that the weaker faction goes away to seek new hunting grounds.

While selection pushes the species as a whole towards a general rise in fitness, selection causes one species to split into two or more distinct, intersterile species only when combined with isolation. When all members of a species form a single interbreeding unit—that is, when no physical barriers restrain genes from traveling, by successive matings, from one end of the species' range to the other—this travel will mix the gene pools of different parts of the range so that all remain more or less uniform. Any local races that arise will never differ so widely from the rest of the species as to become unable to interbreed with them.

In our own species, the range was broken up by mountains, deserts, and oceans, which to hunter-gatherers were practically impassable. In some places, passages from one land to another, which had been open, were cut off by the rise of the sea level with the melting of the Pleistocene ice.

Furthermore, the bands were more or less isolated from one another as a matter of economic necessity. A band of a hundred hunter-gatherers (I call them theratics) needs a range of one to two hundred square kilometers to itself, if it is not to exhaust the local food sources. This isolation hindered the interchange of genes and fostered the divergent evolution of local races towards separate species. But in mankind, the process has never gone far enough to render interracial matings infertile, as it might have in another million years or so of theratic life. Now the overriding tendency in the species is towards greater and greater intermixture and homogenization.

All the human evolution that we know about took place before the rise of civilization. So far as we can tell from bones and art, the first men to till the ground, build cities, smelt metals, and write were not biologically different from us. Since they lived less than 12,000 years ago, we have not been civilized long enough for much evolution to have taken place in that time. Twelve thousand years is less than one percent of the time since the beginning of the Pleistocene, when our ancestors were lowbrows of the *Homo erectus* type.

Genetic drift is the random variation in small interbreeding populations away from the original type. It comes about partly through mutations and partly through the chance loss of types of genes regardless of their adaptive value. For example, if in a large population a certain percentage carry the genes that cause blue eyes, and blue eyes (we may assume) have no selective advantage or disadvantage, the percentage with that gene will remain about the same indefinitely. But if you have a group of twenty people, two of whom carry the blue-eye gene, and those two perish in an accident, then the blue-eye gene will be altogether eliminated from that group's gene pool. Some peculiar traits among small, long-isolated populations, like the high percentage of Rh-negative blood in Basques, may be the result of genetic drift.

I suppose that all my readers have some idea of the workings of heredity. You know that the hereditary mechanism is contained in the chromosomes, which are threadlike particles in each cell of the body (forty-six in each human cell) in nearly identical pairs; that each chromosome is a string of genes, the number running into tens of thousands; that each gene is a complex molecule, which governs, by itself or in combination with others, the growth of one or more parts of the body.

You also have, I am sure, a good idea of the meaning of the terms "dominant" and "recessive." The gene causing brown eyes dominates that causing blue eyes, which is recessive. This means that, when you mate a human being of pure blue-eyed ancestry with one of pure brown-eyed ancestry, all the offspring (with rare exceptions) will be brown-eyed. But when these brown-eyed persons mate with others of similar ancestry, on the average one-quarter of their offspring will be blue-eyed and breed true; one-quarter will be brown-eyed and breed true; and the remaining two-quarters will be brown-eyed but will give mixed offspring in the same ratio of 1:2:1 as their parents did. (There is a great deal more to the story than this, but I am not trying to write a textbook on genetics.)

As for genetic drift, mankind has now become one vast interbreeding unit of about four billion. Travel and migration keep mixing genes from the ends of the Earth, so that genetic drift is no longer effective, save perhaps in small, self-isolated groups like religious cults. Likewise, the loss of isolation on one hand and the development of artificial means of combating extremes of climate on the other—fire, clothes, houses, air conditioning, and vitamin pills—have cancelled any tendency of the different races to continue diverging until they become distinct species.

Racial pride and prejudice can slow the process of interracial mixing but are unlikely to stop it. They did not stop it in India, where over 3,000 years ago the barbarous Aryans set up a caste system with strong religious taboos against intermarriage between castes, to keep their descendants distinct from and dominant over the much darker but more civilized folk they had conquered. Hence today all East Indians, given the same amount of exposure to the sun, have fairly uniform complexions.

Does this mean that mankind will eventually reach one uniform racial type? Such a man would have about 60% Caucasoid ancestry, but half of these "white" ancestors would be "dark whites," like those of India and North Africa. He would be about 30% Mongoloid, and the remaining 10% Negroid. He would be a medium-sized, light-brown, black-haired fellow easily taken for a Mexican or a Polynesian.

It is hard, however, to imagine that enough Caucasoids or Negroids would ever migrate to Mongoloid China to affect seriously the racial makeup of the Chinese. So the main groups are likely to remain distinct, for the foreseeable future, where they now predominate over large areas, such as China, Europe, and sub-Saharan Africa. The main change is likely to be the disappearance by assimilation of little racial pockets and enclaves, such as the Pygmies of the Ituri Forest of Zaïre, the Caucasoid Ainu of northern Japan, or the Mongoloid Kalmuks of the lower Volga.

The big change in human evolution during the last ten thousand years has been the letting up of the pressure of selection, so that the species is no longer visibly improving. In the absence of selection, man has worked himself into an environmental niche like that of the lamp shell and the horseshoe crab, so well adapted to their environments that they have changed but little in many millions of years. Of course, men did this more by changing their environment than by evolving themselves. If the species is going anywhere, it is going downward, because millions now survive and breed despite hereditary weaknesses that would have killed them off in primitive life, and they pass these defects on to their descendants.

But are not millions better off because of improvements in diet, exercise, and medicine? Yes, civilization does have these effects, at least on a sizable minority of the species. But the effects are temporary, because they affect the phenotype, not the genotype.

A man's genotype is his basic hereditary plan, as set by his genes. His phenotype is what he has actually grown up to be, as the thousand natural shocks that flesh is heir to work him over to modify the original plan. One might compare the genotype to the blueprints for an automobile and the phenotype to the actual car after it has been driven a while, with all the mistakes made in the factory, dented fenders, torn upholstery, and other damage.

It used to be thought that by affecting the phenotype you could modify the genotype, as in the story of Jacob and Laban in the thirtieth chapter of Genesis. Jacob had a deal with his father-in-law Laban whereby, when they divided their flocks and herds, Jacob would get all the spotted ones. So crafty Jacob peeled spots on sticks of green wood and set the sticks up in watering troughs where the animals would see them when they came to drink. According to the Good Book, Jacob was thus enabled (by pre-natal influence and inheritance of acquired characteristics) to make away with most of Laban's best stock. Not surprisingly, this act caused hard feelings.

The last important believer in the inheritance of acquired characteristics was that eminent faker Trofim Denisovitch Lysenko. Stalin put Lysenko in charge of Soviet agronomy for fifteen years, with the result that the poor Russians ended up hungrier than ever.

Even the phenotypic effects of civilization are not all to the good. In prosperous lands, millions grow up bigger and healthier because they eat better, but an even larger number suffer from dental troubles caused by too many sweets and soft, rich foods, and obesity from a poor diet and lack of vigorous exercise. Millions become stronger from sports and exercise, but other millions grow up soft and under-muscled from riding in automobiles instead of walking, and watching television instead of playing outdoors. While modern medicine saves millions, tobacco and automobile exhausts inflict lung cancer and emphysema on others.

While any organism is affected by its environment, the effect, with rare exceptions, is on the phenotype, not the genotype. The ordinary environmental factors—food, exercise, sunlight, injuries, and infections—have no effect on the genotypes of future generations.

The one feature of civilization that does have a selective evolutionary effect is the epidemic diseases that arise among civilized folk. The reason they arise is that an epidemic disease needs a certain density of population in order to spread.

When people are as thinly scattered as they are in the hunting-gathering stage, the disease dies out. Consequently, civilized people have greater immunity to such diseases than primitives.

You may have heard how the Amerinds of the Caribbean died out when the Spanish enslaved them. They died, not because they were too proud to live as slaves, but because the Spaniards brought smallpox and measles. From West Africa they also brought Negro slaves and the *Anopheles* and *Aëdes* mosquitos. Anopheles carries malaria, while Aëdes spreads yellow fever. The Indians returned these gifts by giving the Spaniards syphilis and gonorrhea, which also inflicted a devastating mortality when first introduced to Europe.

Selection still works against the most glaring defects, such as juvenile amaurotic idiocy, because the victims die before they can beget progeny. Some other selective influences might, if continued long enough, have an evolutionary effect. If gentlemen really preferred blondes and continued to do so for a few thousand years, this preference might cause a measurable rise in the incidence of blondeness. Celibate priesthoods tend to drain off the most intellectual people and so might in time lower average intelligence.

Polygamy might have the opposite effect by giving rich and successful men more than their due share of the women and thus enabling them to leave more offspring than other men, like some of those medieval sultans and amirs who could muster a whole company or battalion of their own sons.

Evidently none of these forces has been at work long enough to have visible evolutionary effects. Catholic countries like France and Italy, I can assure you from firsthand acquaintance, have not been reduced to drooling idiocy by celibate priesthoods; nor has polygamy yet made supermen of the Arabs or the African Negroids.

Among stone-age primitives, warfare used to have some selective effect by favoring the survival of the strong, alert, agile warrior. But when men began shooting things at one another, from arrows to nukes, survival became more and more a matter of luck. So now I cannot see how war has any particular selective effect.

This brings up the question of atomic war. First, the main effect would be simply to kill vast numbers—perhaps a substantial fraction of the world's population—and leave other multitudes crippled by radiation effects. On the other hand, all we know about these effects indicates that they would be the same as those from other radiation sources like X-rays. Secondly, they would merely speed up, a little, processes that go on all the time.

Hard radiations have effects of two kinds. One is the effects on the victim's body or phenotype, which range all the way from temporary loss of hair to death. The other effects are on heredity, by causing mutations, or sudden hereditary changes. These changes were called "sports" before scientists got around to studying them. The first mutation definitely recorded was a short-legged lamb, born in 1791 in a flock belonging to Seth Wright of Massachusetts.

The kind of mutation that occurs in animals is the gene or point mutation.

Now and then an accident befalls a gene. Some atoms are knocked off, or twisted askew, or an extra atom is added, or the gene's position is changed, or the gene is duplicated or lost. If the cell is an ordinary body cell, such as one from your tongue or your eyeball, nothing happens. But if the cell is a gamete or sex cell, which becomes another individual, the new gene pattern is passed on to that offspring.

One gene in a gamete has little chance of mutating. But with tens of thousands of genes in each cell, mutation is not so rare. Some geneticists think the gamete has about one chance in four of mutating. Since you get one gamete from each parent, you have an almost even chance of being a mutant.

Then why don't we all have two heads? Because most mutations are so small that they can hardly be detected. In fact, there may be a vast number of mutations where the change in the gene has no effect on the resulting organism. When a mutation does have an effect, it may make your eyesight a little keener or dimmer, or your digestion work a little better or worse, or your arteries harden a little sooner or later.

In point of fact, most mutations that have any effect are harmful or destructive. They make our eyes and digestions and arteries worse, not better. Constructive or beneficial mutations cannot be more than a small fraction of one percent of the total.

The bigger the mutation, the smaller its chance of being beneficial. Most drastic mutations are lethal, killing the organism in embryo. The reason is that a gene is an enormously complex little piece of biochemical machinery, delicately adjusted to its task of controlling the growth of some part of the body. To expect a big random change to improve it is like trying to improve your watch by hitting it with a hammer.

Any gene may mutate in many different ways, but some genes mutate more readily than others, and some undergo certain mutations over and over. Thus the mutation responsible for hemophilia occurs about once in 50,000 human births. The hemophilia that Queen Victoria passed on to her descendants in the Russian and Spanish royal families was probably such a mutation.

Although rare, constructive mutations also occur. That is how evolution takes place. In a wild state, organisms with destructive mutations tend to die young, while those with constructive mutations have more than their share of offspring and take the place of those without them. In Europe, the black mutation of several species of black-and-white speckled moth became beneficial around sooty cities like Essen and Manchester, because black moths are harder for hungry birds to see against sooty walls and tree trunks. So the black moths became the dominant form in those areas. Now that the British have cleaned up much of the air of their industrial cities, the black-and-white moths are making a comeback.

We know some of the causes of mutations. Hard radiations, like those from X-rays and atomic explosions, cause some; so does the slight but constant radioactivity from the air and the Earth. About thirty years ago I was prospecting for uranium by airplane in the Adirondack Mountains. Several times I thought I had made a strike; but when I hiked to the spot on foot, I learned that any large exposed mass of granite, such as a crag or a quarry, emits enough radiation to register on a Geiger counter.

As far as evolution is concerned, however, the slow action of this weak natural radioactivity, acting over millennia, causes far more mutations than would the violent bursts of radiation from atomic explosions over a short time. An atomic war would probably not last long, and civilization would suffer far more from the destruction of cities and the breakdown of institutions than from a rise in the mutation rate.

This brings up the question of the safety of nuclear power, which has been a subject of lively dispute ever since the Three Mile Island accident. As far as I can see, the hazard of a meltdown is serious, and the people at the power station were incompetent. But in the worst possible case, the damage from such an accident is of the same order of magnitude as a lot of other hazards that we tolerate, and a lot less than some.

For example, people are killed every year in coal-mine accidents and oil-refinery explosions. We let airplanes fly over densely populated areas despite the fact that every few years a mid-air collision or a structural failure brings one of these machines down on the heads of those below. Tank cars full of dangerous chemicals, some potentially as devastating as radioactive gases, are always running off the track and splitting open. And then there are the exhausts from steam and internal-combustion engines, and acid rain, and . . .

All these sources of risk, including nuclear power, are mere flea-bites compared to the biggest accidental killer of all: the automobile. We regularly kill about 50,000 a year and injure millions, a greater rate of casualties than the Vietcong were ever able to inflict. And most fatal auto accidents could be prevented; I have seen it done.

The cure is simply a jail sentence for all moving violations. In Los Angeles, about 1925, a judge—I think his name was Robinson—had an infant daughter who was run over and killed. So he started giving jail sentences to all speeders. You would be amazed at how carefully and sedately the cars moved along Sunset Boulevard for a while.

Of course, it did not last. The real estate interests, then politically dominant, and the Hearst papers brought pressure to bear and ended the experiment. Similar opposition would probably stop any such effort today. The episode merely confirms the fact that, while most people approve severe penalties for offenses that they do not mean to do, like murder, they do not want drastic punishment for offenses that they themselves expect to commit. Nearly all drivers like to cheat, if only a little, on posted speed limits.

Anyway, to make a great fuss over the dangers of nuclear power, while ignoring all the other hazards mentioned—some of them worthier of fear—is to act like my friend the late Eric Frank Russell. Eric brushed aside the massacres of

Hitler and Stalin but could work up a terrific charge of indignation over some company that marketed a harmful medical drug. It shows a lack of a sense of proportion.

Besides radiations, some chemicals also raise the mutation rate. Among these are mustard gas, some peroxides, ethyl sulfate, formaldehyde (which comes from automobile exhausts), and caffeine (which occurs in coffee and tea). So far as we can tell, though, these mutations are no different from those that have been affecting our lineage all along, clear back to the original protoplasmal primordial atomic globule. Moreover, many defects, like the Down's syndrome (called Mongolism until persons of Mongoloid race raised legitimate objections) and the cleft palate, are due, not to mutations, but to accidents during the formation of the embryo.

There is still a sinister side to mutations. Among wild animals and primitive men, destructives are always arising and being eliminated by selection. But often, harmful mutations are not easily gotten rid of, because most mutations are recessives. Both genes of a gene pair, one from each parent, must belong to the mutated form of the gene before the mutation affects the offspring. If the organism has only one of a pair of genes from a destructive mutant, the creature lives a normal life but can pass on the destructive gene. If the mutation becomes common, the number of those with one of the pair mutated, called heterozygotes, rises until they begin to mate with each other, whereupon a quarter of their offspring develop the harmful trait and die off. Thus the species reaches a balance between the rates of mutation and elimination.

A common mutation causes an organ or a function to be altogether lost. If the organ or function is not needed, the mutation occurs again and again, by simple chance, until the type without the trait becomes the dominant type. This process is called *rudimentation*; it explains eyeless fish and insects dwelling in caves, and why men and elephants have so little hair. It also explains why certain groups of the Negroid race cannot as adults digest milk; after infancy they lose the capacity to make the necessary enzymes.

Conversely, you can be reasonably sure that, when an organism displays a prominent feature, the trait has some use in enabling the species to survive. It used to be thought that the recurved tusks of the male mammoth and the high dorsal fin of the Paleozoic reptiles called pelycosaurs were useless ornaments, created by some ill-understood irregularity of growth. Now these organs are explained more logically: the mammoth's tusks were snow shovels to get at buried food, and the pelycosaur's fin was a heat-control organ.

When selection pressure relaxes, destructive mutations spread through the species unchecked. Every advance in medicine enables more people with defects to live and breed like everyone else. If we have poor vision, we wear eyeglasses; if we have flat feet, we wear arch supports; if we suffer from allergies, we take antihistamines.

It may be significant that male color blindness is only about 1% among Eskimos, Papuans, and Navahos, but over 7% among the long-civilized Chinese,

Europeans, and white Americans, with American Indians and African Negroes in between. Color blindness is less of a handicap to a civilized man than to a primitive hunter.

Well then, why worry, so long as we have eyeglasses, arch supports, and injections for allergies? If destructive mutations would stop at any point, perhaps we could manage indefinitely with modern medicine. But mutations keep right on and would continue even if there were no such thing as nuclear power. If to weak eyes, flat feet, and allergies we add diabetes, anemia, arteriosclerosis, hemophilia, schizophrenia, albinism, early cancer, and a few other destructives, we shall have a pretty wretched person despite future medicine. Our descendants would be no better off for having to work only one day a week if they had to spend all the rest of their time at the clinic having their hereditary, mutation-caused defects patched up. And if civilization broke down only a little, say from a disaster or the exhaustion of a natural resource, such a race of invalids might utterly perish.

Then what can be done about destructive mutation pressure? Ninety-odd years ago, Darwin's cousin Francis Galton started a movement he called "eugenics," to encourage the "fit" to breed and to stop the "unfit" from doing so, by persuasion, isolation, or sterilization. The early eugenists were on the right track but made such wildly inflated claims as to discredit their movement. Moreover many were people of strong racial, national, and class prejudices. If only, they said, breeding were limited to the better sort of people, like us, all crime, vice, and folly would soon be done away with and man made over into a superman.

Now we know that the problem is not so simple. Suppose a man has what looks like a fault. Perhaps it is not really a defect from the point of view of survival, but merely some trait of looks, speech, or habit that we dislike. Even if it is a serious defect, it may or may not be hereditary. Even if it is hereditary, it may not be easy to eliminate. Since most mutations are recessives, if all those manifesting the trait were stopped from breeding, the character would still exist invisibly among a much larger number of heterozygotes. These would produce a new crop of defectives in the next generation.

Simple calculations show that the fewer people who have a recessive gene, the longer it takes to reduce it still further by stopping the homozygotes who show it from breeding. And here we come to the greatest obstacle to eugenic programs: the time scale.

We live in a period when the world is on an equality kick. Many well-meaning people insist that all men are literally equal, and if they are not, it is unfair and undemocratic and we should pretend that they are. Any disagreement with this view is élitism, and you know what a wicked thing that is said to be. Egalitarianism has been carried to the point where, a few years ago, the Kansas Department of Parks and Recreation put on a painting contest for young people. The prize was won by a finger painting by one D. Jim Orang, who turned out to be an orangutan in the Topeka zoo. One need not be an anthropocentric bigot to see that this is carrying equality a bit far.

Such egalitarians decried the idea of eugenics right from the start. It now seems that they were partly right, but for the wrong reason. The trouble with a eugenic program for human beings is the length of the human generation, which is much longer than those of most animals—for instance, about a thousand times that of the geneticists' favorite experimental animal, the vinegar fly *drosophila*. (If you have never seen a *drosophila*, leave the lid of your garbage pail ajar for a couple of days when it contains the remains of fruit, and the vinegar flies will appear in swarms.)

I once calculated how long it would take to reduce the percentage of albinism in the population by preventing all albinos from breeding. It would take 1,450 years to cut the present percentage in half, and 5,000 years more to reduce it by half again, or to a quarter of its present frequency. And no government in history has ever lasted so long, let alone followed a consistent policy for such a time.

That is the weakness of the spermbank for geniuses recently established in California. I do not see that it will do any harm, and if it could be kept going for a few thousand years it might show significant results. But who is going to keep it going for several thousand years? Man has been civilized only for five thousand years.

Moreover, we must remember that, as a result of the chance reshuffling of genes with each generation, the offspring of a pair of geniuses will on the average be more intelligent than average but less so than their parents. Likewise, the offspring of a pair of morons will on the average be more intelligent than their parents but less so than the average of the species. The offspring of parents at the extremes of the curve measuring any variable quality tend to "breed back" towards the average. At the same time, the offspring of average persons now and then turn out to be geniuses or morons, so that the overall percentages of these various types in the species remain the same.

Furthermore, defects like albinism and the rest could not be wholly eliminated, because new mutations would continue to give rise to it. The best we could do would be to keep the percentage down to the frequency it would have in a wild or primitive population.

But couldn't we quickly reduce such common defects as poor eyesight? Look at all the people who have to wear glasses. The trouble here is that "poor eyesight" is not a single genetic fault but the sum of at least a hundred different defects. Each by itself is fairly rare and so would need hundreds of years to reduce even by half.

So eugenics would never make us into supermen in reasonable time. We shall probably never get so godlike a race as some of my fellow writers have imagined, because we can only encourage gene combinations that actually occur. In theory, perhaps we shall some day be able, by using recombinant DNA or something, to tailor human chromosomes so as to eliminate genes with destructive mutations. This, if it ever happens, is probably a long way down the road; but I have been wrong before by being too conservative in trying to guess when some future advance would materialize.

Critics of eugenics used to say that we could never agree on a plan for our superman. Should he have the head of Apollo, the body of Hercules, or the brain of Einstein? Now we know that beneficial mutations are so rare that there is no competition among them. Any real improvement in health, strength, intelligence, longevity, or disposition would be welcome. In practice, eugenics is more like running as fast as we can to stay in the same place.

Others say that we should instead practice euthenics—improving the conditions of life for the masses. But there is no conflict between these aims. Once people get plenty to eat and as much education as they are able or willing to absorb, they are unlikely to advance further with their present genetic equipment. Millions are in this happy position today—a vastly greater number, both relatively and absolutely, than was the case a few centuries ago—but that fact does not stop them from acting in foolish, irrational, self-destructive ways. You need only read the newspapers to see that. Anyone who thinks such advances in the level of human conduct can be brought about by some magical change in educational methods, political organization, or economic system is kidding himself.

Destructive mutation pressure, however, continues, and the only way now known to cope with it is eugenics. The prospects of eugenics are not quite so dim as might appear. Sometimes a recessive gene can be detected by small, telltale effects. In such a case it could be eliminated much sooner than if we knew about it only in homozygous form. But such a program would need stabler governments than now exist, to continue the program long enough to do any good.

Finally, both selection and mutation pressure work very slowly. While we shall eventually have to face the problem of destructive mutation pressure, a delay of a century or two will not be fatal and will allow the discovery of enough additional knowledge to do the job right.

In the meantime, the problem of, not the quality, but the *quantity* of human beings will be much more urgent. Some years ago I calculated that at the then rate of increase in the world's population, in about 1,400 years there would be one human being for every two square feet of the Earth's land surface. Since the average person takes up that much space when standing, that is literally standing room only. Something obviously will have to give long before that point, and what is likely to give is mass starvation on a scale we cannot even imagine. We had an ominous example a few years ago in the Sahel famine in Africa.

Those who talk of relieving population pressure by shipping people off to other planets or space stations are dreaming. Such a project is about as realistic as trying to beat a rhinoceros to death with a fly swatter. It is possible, at great expense, to maintain small groups of scientific specialists on the moon, or on Mars, or in a space station. Such people will discover (and have already discovered) many interesting things about the evolution of the universe.

The conditions of life in such places, however, would be much more exacting than life in Antarctica, where such colonies are maintained by leading powers now. But we don't see any massive migration from overpopulated lands to Antarctica. The annual increase in the population of Asia alone (figures are uncertain) has been estimated at 14,000,000; and where on the Earth or off it could you put so many, not just once but year after year? Anti-abortionists may take note. They generally avoid the subject, as if it would go away if ignored long enough.

So, about man's biological future, we can be sure that it will be crowded. It looks as if, for the near future, the numbers are increasing rapidly, while the quality is slowly declining. Let us hope that both trends can be halted and perhaps even reversed while there is still time to do so.

Part Four

THE SEARCH FOR

EXTRATERRESTRIAL

INTELLIGENCE

David Brin, Ph.D.

XENOLOGY: THE NEW SCIENCE OF ASKING "WHO'S OUT THERE?"

MAY 1983

SPECULATION about extraterrestrial intelligence has had a long history, which has been particularly turbulent of late. Not long ago, it was strictly the province of science fiction and fantasy; most "respectable scientists" were highly skeptical about the likelihood of life on other worlds and even more skeptical about our being able to interact with it even if it existed. Now, things have changed to the point where most scientists suspect that life, even intelligent life, should be *common*, and that interstellar communication and travel should be quite feasible. So, where is everybody? That question, often referred to as "The Fermi Paradox," has become one of the hottest topics of speculation for scientists in several fields.

David Brin is one of them. With a 1981 Ph.D. in space physics from the University of California at San Diego, he started off to make a career in that field—but has been largely sidetracked from it by his spectacular success as a science fiction writer. Multiple Hugo and Nebula awards for such stories as *Startide Rising* and *The Uplift War* have encouraged him to devote more time to writing, but he remains very actively interested in the frontiers of science.

This article, by the way, is a good example of *Analog* readers' active interest. In it, Brin poses a challenge to readers to offer their own ideas on the subject—and their responses led to a follow-up article in July 1985. ∎

IN the early 1960s, while the world was entranced by the spectacle of human beings hurled into "outer space" in rocket ships, a series of philosophical earthquakes shook the sedate field of astronomy. Just when the skies were beginning to seem known and familiar, all at once things changed. Stellar astronomers suddenly faced unsettling data from new classes of objects called "quasars" and "radio galaxies." There were disturbing theories about so-called "black holes." Even those who had long studied planets now found their comfortable domains invaded by geologists and meteorologists, who weren't at all shy about moving into the new territory.

It was no coincidence that all of this happened just as the Space Race was getting under way. New instruments and techniques often lead to upheavals in a science.

Still, the greatest intellectual challenge to the world view of modern astronomers came in the early '60s not because of new space probes, telescopes, and computers, but because of an idea.

Starting in 1959, with a classic paper by Cocconi and Morrison, a series of articles and books were published with titles like *Interstellar Communication, Habitable Planets for Man* and, in 1966, a major work entitled *Intelligent Life in the Universe* by Iosef Shmuelovich Shklovskii and Carl Sagan. With these studies, astronomers began to concern themselves with life itself.

The early '60s were pivotal for the field of "exobiology"—extraterrestrial biology—and especially the sub-branch that dealt with intelligent life, "xenology." For the first time it was legitimate for leading scientists to publicly consider the possibility of contact with intelligent species off of the planet Earth.

Of course, a lot of thought had gone into the subject previously, on the pages of science fiction novels and magazines. Many of the private discussions between authors such as Sagan, radio astronomer Frank Drake, and Rand Corporation scientist Stephen Dole grew out of ideas germinated by the likes of Clarke, Asimov, and Clement during the '30s and '40s. But prior to the publication of *Intelligent Life in the Universe* the number of "respectable" papers on the subject printed in the West could be counted on one's fingers and toes.

In the Soviet Union extraterrestrial intelligence was not only considered possible, but was required by Leninist dogma. (It was assumed dialectically impossible that any advanced intelligence could be anything but socialist, of course.) When scientist I.S. Shklovskii wrote "Universe, Life, Mind" in 1962, his thoughts were widely popularized, and extracts were reprinted in major Soviet scientific journals.

In the West it took more time for scientific speculation about the distribution of life in the cosmos to become acceptable. A tradition of skepticism and rigor kept western science relatively safe from scientio-religions such as Lysenkoism, which caused so much harm in Russia. But the same attitudes made it hard for those interested in the possibility of alien life forms to bring up the subject in scientific gatherings without being criticized for "playing with science fiction."

The older scientists who dished out the ridicule shouldn't be blamed too harshly. In squelching early discussions of exobiology, they may have been overreacting to the excesses of earlier enthusiasts, such as Percival Lowell, the astronomer who convinced millions that there were living Martians and "networks of canals" on the red planet.

But with the arrival of the space age, resistance to "science-fictional" ideas was dealt a fatal blow. Those who had declared that "spaceships" belonged only in comic books were caught flat-footed. A new generation of scientists brought exobiological speculation out of the fringes and onto the pages of respectable journals. These men and women, who had proven their scientific credentials with solid research, came from an age group which didn't consider "science fiction" a dirty word. Most of them had cut their teeth on the stuff.

The first time I witnessed the subject of extraterrestrial intelligence brought up at a scientific seminar was at a Wednesday Caltech colloquium in 1968. The speaker remarked on the remote possibility that pulsars might be beacons of an advanced civilization. They were, after all, several thousand times more regular in their repetitive "beepings" than any other astronomical radio source ever discovered.

The speaker was only partly serious, but sides were quickly taken, and it was soon very clear that most of those with tenure didn't like this kind of talk at all.

Attitudes were changing very rapidly during those years. A few years later some of those who were the angriest in 1968 applauded the loudest when Carl Sagan unveiled the gold plaque that was to be placed upon Pioneer 10, the first human artifact to be launched on a trajectory out of the solar system.

Today that plaque is famous. It, and those which followed on Pioneer 11 and the Voyagers, depict the nude figures of a woman and a man, an arm raised in greeting, a schematic of the planets of our system, and a rayed pattern of lines and binary dots representing the most prominent pulsars detectable from Earth. The pulsar map should enable any distant beings who recover the spacecraft to trace its point of origin within a light year in space, and its launch date to within six months.

Shortly thereafter respectable scientists were discussing, not whether extraterrestrial intelligences exist, but how to go about listening for signals from our nearest neighbors! Small (very small) amounts of public money were allocated to adapting radio telescopes for the search.

If the first revolution in the nascent field of xenology came on the pages of science fiction pulps of the '30s and '40s, the Second Xenological Revolution took place in the '60s, when scientists in large numbers began asking, "Where are they?"

Anyone interested in the possibilities of life outside the Earth should certainly read *Intelligent Life in the Universe*. Although some of its science is dated, it remains the classic in the field. Still, to a veteran reader of SF, *Intelligent Life . . .* may seem overly tame and conservative. For instance, the authors barely mentioned the possibility of *travel* between the stars. To investigators of that time, it seemed pointless to discuss interstellar colonization.

Science fiction has long used, as furniture, ships which bypass relativity. But to early xenologists it was dangerous enough talking about alien life forms, without risking one's scientific reputation talking about "hyperspace warpdrive" and the like. Shortcuts may lend SF a lot of pizzazz and spawn stories about galactic empires, but Einstein's speed limit dominates serious talk about life in the universe.

The gulfs separating the stars are vast. And during the '60s it seemed unlikely that even the modest velocities allowed under Einstein's edicts could ever be reached economically.

Thus the first era of modern scientific xenology (from 1959 to about 1972) dealt with the possibility of intelligent life springing up in isolation—here and there on fertile planets scattered across the sky—islands of intelligence separated from one another by vast distances and for all time.

THE AGE OF INNOCENCE

What could early students of this new science say about extraterrestrial life forms? No matter how daring, they were faced with one major limitation: a near total lack of data. The only known case of intelligent life is here on Earth. Until the Viking mission, some held onto Percival Lowell's dreams for Mars. Now the evidence seems to weigh against finding even microbes there. Putting aside, for the moment, speculations about dolphins, whales, and gorillas, it's pretty hard to extrapolate a graph from only one data point.

Still, three scientific discoveries and one useful philosophical tool gave researchers the courage to make crude estimates about the distribution of life amongst the stars.

The first discovery came when it was found almost ridiculously easy to make amino acids, and other precursors to living matter, from abundant molecules such as methane, ammonia, and cyanogen. Stanley Miller subjected a water solution of these substances to electrical discharge and ultraviolet radiation and got an organic "soup" in short order. Leslie Orgel of the Salk Institute accomplished the same thing by a freezing process. The high pressures of ice formation not only gave up amino acids, but the purine adenine as well. (Adenine is one of the four building blocks of DNA, and is the core of ATP, adenosine tri-phosphate, which controls the energy economy of the living cell.)

Xenology: The New Science of Asking "Who's Out There?" / 143

So many mechanisms have been found that can change crude precursors into "biological" molecules, that organic activity seems almost an automatic consequence of the distribution of chemical elements in the universe today.

The second major discovery supports this point of view. During the last two decades, radio astronomers—listening to narrow emission lines from interstellar space—have discovered great clouds of complex molecules: ethylene, formaldehyde, ethyl alcohol; some even claim evidence for—you guessed it—adenine.

(Astronomer and science fiction author Sir Fred Hoyle, looking at starlight scattered from interstellar dust, thinks that the dust itself may actually be something akin to bacteria . . . living cells about a micron in size, in diffuse colonies spanning light years and outmassing suns. It's an extravagant speculation, but fun to think about.)

It's clear, then, from basic chemistry and radio astronomy, that the basic materials for life are out there. What about the right environments? We have to assume, until we have reason to think otherwise, that complex life must grow and evolve to intelligence on planets orbiting stable stars. Are there other "nursery worlds" like the Earth?

There are plenty of stable, long-lived, G-type dwarf stars like the sun out there . . . about 6% of the galaxy's several hundred billion stars. Are there planets circling many of them?

The data are still poor. It's hoped that the Space Telescope will tell us more about the companions of nearby stars. Some scientists think there is good evidence that at least one of our neighbors, Barnard's Star, has possibly two dark companions a bit more massive than Jupiter.

We do know that F, K, and G-type dwarf stars rotate much more slowly than larger, hotter stars. The sun contains 99.9% of the matter in the solar system, yet it has only 0.5% of the angular momentum. The rest is distributed among the planets of the solar system, especially Jupiter. Most astronomers believe that those slowly rotating stars which aren't members of multiple star systems have to possess dark companions that were used to "dump off" excess angular momentum early during star formation. Recent models of gas cloud condensation tend to support this belief.

We've covered three discoveries, then, that help us believe that it's reasonable to talk about life outside the Earth. What is that "philosophical tool" we mentioned that caps the legitimacy of xenology? It is sometimes called the "cosmological principle," or the "assumption of mediocrity."

Since Copernicus, astronomy has been a series of lessons in humility, all leading to the conclusion that "there is nothing special about where and when we are." First the Earth was displaced from the center of the solar system, then the Sun became a nondescript traveller in orbit about the rim of the galaxy. The galaxy became merely one island universe among billions, and the universe seems to have no "middle" at all.

The cosmological principle tells us we should avoid the temptation to think that there's anything unique about the Earth in space, time, or situation. It is the major philosophical underpinning for the new study of xenology. It forces even the most cantankerously conservative astronomer to admit that someone, somewhere, might be peering up at HIS stars, among which insignificant motes is our own sun.

If xenology has some justification, then, where did the first generation of scientific xenologists get their numbers? How did they estimate the population of our galaxy . . . or the probable distance to our nearest neighbors?

The Drake Formula is the most popular way to guess at the possible distribution of technological species. It was invented by Frank Drake when he was at the Arecibo National Radio Observatory. It remains the most widely accepted tool for xenological speculation.

Let N = the current number of technological civilizations in the Galaxy. Then,

$$N = R\, P\, n(e)\, f(1)\, f(i)\, f(c)\, L$$

Here R is the average rate of production of suitable stars since the formation of the galaxy, approximately one per year. (The current rate is slower. R is an average that includes the burst of star creation early in the galaxy's history.) P is the fraction of stars which are accompanied by stably orbiting planets. Factor n(e) is the average number of planets per system which have the requisite conditions to support life.

The other factors include f(1), the fraction of these congenial planets on which life actually occurs, f(i), the fraction of these on which "intelligence" appears, f(c), the fraction of intelligent species that attain technological civilizations, and L, the average lifespan of such a species.

For what then seemed fairly good reasons, Sagan and others chose to assign P and n(e) each values near 1. These guesses, within an order of magnitude, don't seem to conflict with what we now know about planets.

For purposes of discussion it was assumed that congenial planets normally develop life [f(1) = 1], that about a tenth of the planets with life evolve intelligence [f(i) = 0.1], and that about a tenth of the latter will see technological civilizations [f(c) = 0.1]. In other words, a likely planet will contribute roughly 0.01 technological races during its history.

A complete discussion of the Drake equation can be found in books and in many recent technical articles. (Some references for the interested reader are given at the end of this article.) There are reasons to believe that the equation is, in fact, short about three factors. But suffice it here to say that the best guesses, with plenty of up-and-down leeway in every parameter, led Sagan and others a decade ago to a rough estimate,

$$N = 0.01\, L$$

This meant the average lifespan of technological races would determine the number present in the galaxy at any time. If self-destruction is the common fate of "civilized" species, then there might be no more than a handful of them in the Milky Way at a given moment, separated by vast tracts of silent starscape. If, on the other hand, a reasonable fraction of races live a long time, the galaxy might be teeming with life.

Cameron, von Hoerner, Shklovskii, and Sagan all guessed at L, allowing for various ways in which a culture might end. Generally, their results suggested that the number of civilizations in the galaxy might be on the order of one million, most of them long-lived and patient species. The numbers giving rise to this estimate were a bit arbitrary, but not unreasonable.

If the planets of a million stars held sophont races, then about 0.001% of all eligible stars in the galaxy would be inhabited by thinking beings. The average distance separating these islands of technology would be on the order of several hundred light years—a gap which seemed unbridgable corporeally, but easily crossed by radio waves.

This was the state of affairs in the early '70s. With interstellar travel virtually ruled out, the accepted model depicted isolated motes of intelligence separated by sterile tracts of space.

These speculations led to CYCLOPS, OZMA, SETI, and CETI. The search for extraterrestrial life was born. The radio astronomers who slapped together borrowed time and equipment to scan the sky were hopeful, and numerous articles about their endeavors came out on the pages of magazines such as the one you hold.

We could take up several articles just talking about SETI. The early arguments over search strategy are fascinating reading. What kinds of antennae would be best suited for the job? Would extraterrestrial intelligent species (ETIS) transmit in the "water hole" frequencies? Should we transmit our own messages, or just wait and listen? If we wait, should we let our nearest neighbors get their first impressions of us from DEW-line radar and "I Love Lucy"?

Extraterrestrials might not use radio for long-range communication. If lasers carried their traffic, we might not be able to eavesdrop on interstellar conversations.

Even if we can't tap long-distance calls, though, we might still listen for leakage from a planet's commercial radio network . . . or search for a beacon . . . a signal *meant* to be picked up by new radio-using species like ourselves.

Many papers came out during the early '70s suggesting that advanced extrasolar sophonts would likely broadcast the interstellar equivalent of "Sesame Street," to help younger species (like us) pass over their initial dilemma of survival or selfdestruction. The reasoning went that it would be in the older species' interest to help its younger neighbors live long enough to get a decent conversation going.

The first formal search for extraterrestrial intelligence came when Frank Drake and his associates looked at the two nearest candidates, the two sunlike stars that lie within twelve light years and are not members of multiple star systems. Drake's team found nothing but star noise coming from the K2 dwarf, epsilon Eridani. Then they turned their telescope to the star Tau Ceti.

And lo! They heard something! For a brief instant they felt a thrill, as modulated signals came down the cable, obviously of intelligent origin! But then, as the telescope settled down, the "signal" faded away, never to return. They soon concluded that the signal was indeed coded noise from the nearest civilization—some commercial traffic in nearby Milford, Massachussetts!

Undaunted, Drake and others expanded the search. The telescopes turned and scanned. Nothing was found. The Russians joined the search enthusiastically. They reported only negative results.

No problem, astronomers suggested. Any advanced species wouldn't waste energy broadcasting over the entire bandwidth of, say, the hydrogen 21-centimeter line. To conserve power, and to attain a high signal-to-noise ratio, they would modulate over a very narrow band. Just wait, they suggested, until we can develop fineband simultaneous multichannel analyzers!

Yet the second and third generations of eavesdropping devices have come up with nothing.

True, still better instruments are planned. The money and time spent in the search have been insignificant compared with the potential rewards, which might include clues to the very survival of the human race. (There is a battle under way as this is being written, to restore the piddling $2-million appropriation for SETI, which recently was "proxmired" to death.)

Still, just one decade ago some of the radio-xenologists were talking as if they expected to be cracking codes in short order.

Now a few even glumly propose that no one is "out there" after all . . . at least not in our vicinity.

How can this be? If we've been at the search for less than fifteen years, using spare time and borrowed equipment, how could anyone expect success so soon? Sure, it'd be nice to find neighbors twelve or twenty light years away; you could hold a "conversation" within one person's lifespan, for example. But, according to most calculations using the Drake Formula, the average distance between technological civilizations might be a few hundred light years. There are well over a million stars in a sphere a hundred parsecs across. It would take some time to search even the most likely of these, choosing only those radio bands we guess to be the best (not knowing whether our idea of "best" is universal).

Two hundred light years makes "conversation" a little more difficult. But a "Sesame Street" beacon would be just as useful as ever at that range. Just knowing extraterrestrials *existed* might profoundly boost *Homo sap*'s sagging morale.

It seems like we are presenting an argument to restore that appropriation from Congress, not laying out a case for doom and gloom. It only appears to be a matter of time and effort. Success, in the long run, seems assured to the persistent.

What has changed then? What has caused this spreading anxiety?

It's not the sort of thing one would expect to be a cause for pessimism. At first hearing it sounds like very good news.

Starships are possible—

THE THIRD ERA OF XENOLOGY

The Third Xenological Revolution began sometime in the mid-'70s, when several prominent scientists challenged the conventional wisdom that intelligent life arises upon isolated islands, forever separated by the wide gulfs of interstellar space. Sanger, Bracewell, Forward, Bussard, and others demonstrated that it's possible to build spaceships to cross the emptiness between the stars. No "magic" is needed. It isn't necessary to repudiate Einstein. Whether by light sail or by antimatter rocket, humanity may be launching starships within a few centuries.

These "starships" would be nothing like the good old *Enterprise*. Limited to possibly a tenth of the speed of light, they could not travel terribly far by interstellar standards. But clearly they could carry people, possibly living several generations in transit. The "slowboat" generation-ship of science fiction fame has been mathematically vindicated.

This is bad news?

Of course not. But the possibility of starships places a new and awesome burden on xenology. It presents us with a paradox that is very difficult to overcome.

What would WE do if we had starships? If both history and literature tell us anything, we would look around for nice real estate and start colonizing. In fact, we wouldn't even need to find nice planets; stable stars with asteroid belts would do. Our own "belters" might by then prefer such virgin territory to "dirty planets," anyway.

Once the new colonies reached a high level of industry, say in a few hundred years or so, what would they do? Why, they'd send out more colony ships, of course. It seems obvious to almost anyone holding a magazine like this one.

Imagine a sphere of human settlement slowly expanding through space. How long would it take for colonies to be planted 300 light years from Earth? Even limiting ship speed to a tenth of the speed of light, and allowing each colony plenty of time to industrialize? Ten thousand years? Thirty thousand years?

Mankind has hardly changed at all, physically, in the last thirty thousand years. If we make a few social advances and avoid self-destruction, we should be able to fulfill the above scenario.

And why shouldn't anyone else? If this sort of expansion can occur once, why not for each of the million sophont races we calculated earlier? In well under 100,000 years the 200 light-year "average spacing" between races would be filled up!

Recent calculations by Eric Jones of Los Alamos Laboratories indicate that the scenario we have just described, of a slowly expanding sphere of settled solar systems, could fill the entire galaxy within sixty million years. It's not unreasonable to imagine at least one out of a million civilized races living that long. So why do we see no signs that the Earth has been colonized in the last sixty million years?

Why have we picked up no radio signals, when the stars should be humming with information and commerce?

Where are they?

This question marks the first traumatic awakening of the new science of xenology. It marks the end of a very short period of innocence. Starting around 1975 and building toward the present, the Third Xenological Revolution commenced. The dust has not yet settled, but one thing is clear. Some of our assumptions are wrong. The universe might turn out to be considerably more complicated than the scientist optimists of the late '60s at first thought.

Of course, science fiction writers and readers could have told them that all along.

THE GREAT SILENCE

The third revolution in xenology came with the realization that space *should* be filled with intelligent life. There appears to be no excuse any longer for the failure of SETI.

Indeed, why hasn't the Earth itself been colonized! The question, "Where are they?" might better be put, "Why aren't they HERE?" The quandary can be called the "Mystery of the Great Silence."

We see no evidence for ancient alien cities in the Earth's crust. Venus and Mars apparently never were terra-formed, though many now think we could tackle the job in a few centuries. The asteroids of the solar system appear to be untouched.

Most significantly, the Earth, until less than a billion years ago, was populated for two billion years by only primitive prokaryotic organisms. A visiting starship need not have landed colonists. All they'd have had to do is be careless with their garbage or latrine, and the history of the Earth would be totally different.

It certainly looks as though we've been alone a very long time.

There have been several imaginative suggestions to explain the Great Silence. At the end of this article we'll compile a partial list.

Dr. Eric Jones, Dr. Frank Tipler, and Dr. Michael Hart all think it means that the earlier calculations of the probabilities of intelligent life were greatly overoptimistic. They suggest that the apparent absence of ETIS simply means that this part of the galaxy is uninhabited . . . that no race has got out there ahead of us to make an impact by colonization. Their "Uniqueness Hypothesis" implies that some or all of the factors f(1,i,c) in the Drake equation are really very small. For instance, some contend that intelligence such as ours is an evolutionary fluke.

Dr. Thomas Kuiper of JPL has presented strong arguments in refutation, showing that convergent evolution has happened frequently on Earth and might well occur elsewhere.

Dr. John Ball has dredged up the science fictional idea that the Earth is a "zoo" or wildlife preserve, and that extraterrestrials are already here, observing us. There are many variants to this concept, including "quarantine" (ETIS awaiting humanity's social maturity), a non-interference "Prime Directive," and many others. All imply we should add to the Drake equation a factor to account for ETIS purposely avoiding contact.

Contact optimists, such as William Newman of Princeton and Carl Sagan of

Cornell, have tried to make excuses for the extraterrestrials. In a recent paper Newman and Sagan suggested that truly advanced cultures would practice zero population growth and thus feel less pressure to expand into virgin territory. The rate of "galaxy-filling" calculated under their extremely conservative assumptions is slow enough to make it barely possible that the nearest expanding space-faring race simply has not reached us yet.

Sagan and Newman further propose that techniques of life extension—immortality—would make individuals of a race very conservative. If a passion for risk-avoidance took hold, a species' rate of expansion, "V," could drop to nil.

Might a race naturally graduate to other interests after a certain amount of time? Science fiction is filled with possibilities, from extra dimensions to realms of the mind far more attractive than drifting through space and clearing land on some new world. Such "maturity stages" would affect "L" in the Drake equation, as well as the velocity of expansion.

Our assumptions for f(1) might be too high. Although the precursors of life —sugars, amino acids, nucleic acids—seem likely to be about as common as stardust, it's possible that the next steps to life might be much, much harder to reach, requiring some rare catalyst to set the process off.

From physics and SF comes the dreadful idea of "deadly probes." Saberhagen's "Berserkers" might make life rare if some technological civilization accidentally let loose something so monstrous. Gregory Benford's variant on the idea is hardly more optimistic. A particularly paranoid advanced species might not want any potential competition to rise up elsewhere. Self-replicating autonomous probes might be sent out to reproduce and fill the galaxy. Whenever new radio traffic indicates that new sentients are loose, these preprogrammed probes would home in on the signals with powerful bombs and stop the infection before it spreads.

It's already too late to call back the spherical wave of "I Love Lucy," etc., that's already spreading through nearby space.

All of the hypotheses given above have their problems. Some seem to contradict the best knowledge we have in the field. Others, like the "zoo" theory, are almost innately untestable.

What we hope to do in this series is compile a list of these possibilities, with the aid of the readers of this magazine. Ideas which *do not conflict* with known facts about the universe will be welcome. When the list of possibilities is published, we will acknowledge those ideas which are original and seem to have merit.

A reading list will be provided at the end of this installment, in order to assist those serious about finding out more about the subject.

I will start things off by talking about a few hypotheses that the xenologist speculators have mostly passed up. Some are a bit frightening.

THE FATE OF "NURSERY WORLDS"

In the Drake Formula the combined factor f(i,c)—the fraction of life-planets on which intelligence and technology eventually evolve—is generally assigned a value of about one in 100. The xenologists who put forward the "one-percent" argument support it by citing the apparent fact that it took four billion years for the Earth to give rise to merely one technological race. This is almost half of the viable life span of the planet. Intelligent life would seem to be a rare and wonderful thing.

But is this assumption tenable? It appears to be the weakest link in the chain of logic.

Let's consider the life cycle of a "Nursery World," a planet with a stable biosphere in which the slow evolution to intelligence can take place.

Evolution appears to have proceeded gradually at first and then at an accelerating pace for over three billion years. Except for (maybe) the introduction of sex, and later of flowering plants, there is no evidence in the fossil record to support the idea that the Earth was ever suddenly invaded by extraterrestrials who, "with kith and kine," introduced advanced flora and fauna. The Great Silence seems, at first glance, to have stretched through the entire Paleozoic.

If we assume the Earth lay untampered with until at least the time of the Jurassic, we can guess that it takes about three billion years for life on a Nursery World to evolve to a level of complexity that makes intelligence feasible.

What if humanity suddenly vanished? Would it take another three billion years for intelligence once again to arise on Earth? If so, it's reasonable to accept the guess that the number of technological species to erupt per habitable planet is of order less than one.

But *Homo sapiens* is not the only species to have benefitted from three billion years of evolution. Today's German cockroach may look a lot like his distant ancestors, but he has accumulated many little tricks his cousins in the Triassic never heard of. The size of genome of the raccoon and wolf is hardly smaller than that of man.

Consider what's happened since the Cretaceous-Tertiary Catastrophe approximately sixty-five million years ago—the disaster which wiped out, over a period of a few hundred thousand years, almost every species of land animal whose adults massed more than forty kilos.

The creatures whose descendants went on to dominate the planet were small mammals: the early equivalents of mice, lemurs, and tree shrews. These humble animals expanded and diversified to fill all of the ecological niches left vacant by the demise of the large reptiles. We are among their descendants.

In spite of the present arms race, man still lacks the ability to exterminate mice, although he will probably soon be able to do an efficient job on himself. The sudden demise of this star system's current technological race would not finish off the Earth as a nursery. If "mice" did it once, they could probably do it again.

We are led to suggest that suitable worlds must pass through long initial "fallow" periods before attaining a level of biological sophistication ripe for intelligence. Afterwards (as proposed by John Gribbin in the December 7, 1981 *Analog*) such planets should be able to produce sophont species at fairly short intervals, *depending upon the time needed to recover from the damage done by the previous sentient race.*

The interval between the Cretaceous Catastrophe and the present is a reasonable estimate for the time it takes to build a civilized race, once small and sturdy creatures have reached a high level of sophistication.

COLONIZATION ECO-DISASTERS

Let's go back to that expanding spacefaring species we were talking about earlier. Remember, calculations show that it might take as little as sixty million years for such a race to fill the galaxy. A question seldom asked by science fiction authors who write about colonization is, "What happens to the colonized planets?"

Unless the settlers leave large parts of their worlds fallow in wilderness preserves, or engage in "Uplift" bio-engineering of local higher animals, their mere presence is likely to prevent the appearance of local sentient species. The cycle of production of intelligent species on a planet is probably delayed indefinitely by an active technological settlement. A world is not likely to serve as a useful nursery of intelligence so long as it is occupied by a spacefaring race.

When the tenants finally do vacate (or die off), the recovery time required before another generation of tool-users evolves will depend on the way the settlers treated their adopted world. The more savage the exploitation of a colony planet, the more severe will be the thinning of the local biosphere. Our own technological civilization has markedly simplified ecological networks on Earth even where efforts have been made to preserve wilderness. In general, higher life forms, more delicate and dependent upon complex environments than smaller creatures, go first.

When settlers finally do step aside—by attrition, disaster, exodus, or whatever—ecological recycling can resume, but recovery and regeneration of intelligence will take much more time, the longer a technological race occupied the planet.

EXPANSION SHELLS

It is generally assumed that a spacefaring race will expand into the galaxy because of either raw curiosity or population pressure. Either way, it's clear that the expansion soon becomes spherelike, with only the most recently settled worlds having much opportunity to seek new planets. For a race limited to slow-boat technology, colonization will take place only in a thin shell surrounding an older, settled region within.

If population pressure is the primary motive for expansion, we have to wonder at the fate of the long-occupied worlds in the interior of the settled sphere, especially those near Home planet. The words "population pressure" themselves suggest the likely fate of these worlds.

Consider the settlement of Polynesia from roughly 1500 B.C. to about 800 A.D. The island-hopping analogy with interstellar exploration and colonization is apt up to a point. Jones borrowed growth and emigration rates for his model of interstellar settlement from Polynesian history. The intrepid Polynesian example is used as testimony to the likely success and viability of "star-hopping" colonization ventures.

Polynesia may, indeed, be representative of interstellar settlement, but not in a pleasant sense. The Hollywood image of island life is paradisical, but Polynesian cultures were subject to regular cycles of extreme overpopulation controlled by bloody culling of the adult male population, in war or ritual. There are many stories of islands whose men were almost wiped out: sometimes by internal strife, sometimes by invading males from other islands far away.

Meanwhile, introduction of domestic animals disrupted island ecosystems. Many native species were wiped out.

The most severe example is the island of Rapa Nui, also called Isla de Pasqua, or Easter Island. Isolated thousands of miles from its nearest neighbors, it was as much like an interstellar colony as any place in human history, when it was settled around 800 A.D. Mankind may devoutly hope to do better when finally embarked to the stars.

The Pasquans utterly destroyed the virgin ecosystem of Rapa Nui in a few generations, ravaging the forest until only banana trees were left. When no wood remained for houses or boats, they had to abandon the sea and its resources, along with all possibility of escape or trade. What remained was native rock—which they carved into hauntingly desolate images—and warfare.

When Europeans arrived, the natives of Rapa Nui had just about destroyed themselves.

Assume a settled sphere of expansion by an extraterrestrial intelligent species. What of the inner systems, *within* the sphere? The Polynesian example suggests a dismal image of increasing competition for dwindling resources with no escape valve for excess population, since all surrounding systems are in similar straits.

What happens to these inner worlds? They probably don't go looking to conquer their neighbors. Interstellar warfare seems to be a frightfully expensive proposition. Conflict arising from population pressure is far more likely to be local, consisting of struggles for resources within each planetary system.

In an old settled system all available asteroids would long have been turned into habitats. Safe inner orbits with unhindered access to solar power would be at a premium.

Even the most efficient space structures will require frequent replenishment of volatile substances—gases such as oxygen, hydrogen, and nitrogen. Comets might supply part of this need, but terrestroid planets would be closer and rich in the desired light elements.

One might expect to see a profound cultural split between those living on planetary surfaces and those in space. Competition and misunderstandings might tempt the space dwellers to take advantage of their superior position to dominate their planet-bound cousins. It would be simple to bombard the cities on a planet's surface with redirected asteroids until civilization there was obliterated. Factor L clearly falls in such a case.

(The space-borne, long divorced from any attachment to planetary life, might even see a terrestroid planet as a likely source of building materials! It wouldn't be beyond their ability to pulverize a world such as the Earth by arranging planetary collisions. This would certainly affect not only L, but also n(e), the number of *planets* on which life can evolve!)

In any event, the innocent higher animals suffer in the crossfire.

ANOTHER EXPLANATION FOR THE CRETACEOUS CATASTROPHE

Let's return briefly to the episode about sixty-five million years ago known as the Cretaceous-Tertiary Catastrophe. There were, at that time, many advanced species of reptiles. The best candidate amongst these for a species possibly ripe for development toward tool-using might have been *Saurornithoides*, a mid-sized bipedal carnivore with the highest brain-to-body mass ratio of any reptile, approximately matching that of modern baboons. While there is no reason to think that this creature was particularly intelligent, he filled an ecological niche that might have been rigorous enough to encourage his glimmering abilities.

But *Saurornithoides* died out along with virtually all of the other great reptiles during a relatively brief period by geological standards.

If the demise of the dinosaurs puzzles paleontologists, the problem has been even worse for the marine biologists. The dinosaurs, at least, took as long as a few million years to die out. The tiny sea microorganisms experienced a greater catastrophe. Over half of the species of phytoplankton went extinct within about one year!

The latter mystery, at least, now appears solved. Recent deep-core drillings have uncovered thin layers of clay rich in exotic elements, including iridium (up to 25 times normal abundance of some isotopes), at sedimentary levels associated with the end of the Cretaceous. Discoveries in locations as diverse as Italy and New Mexico all seem to correlate a sudden invasion of strange dust with the equally sudden disappearance of many classes of oceanic microorganisms. Scientists now conclude that a major meteorite impact kicked up a great pall of dust which severely altered weather patterns, resulting in mass extinction by starvation when photosynthesis was interrupted.

(The meteorite explanation of the Cretaceous Catastrophe was also discussed by John Gribbin in the December 7, 1981 issue of *Analog*.)

For the marine creatures this seems sufficient, but don't forget the dinosaurs were *already* dying out before this bombardment, starting with the greatest behemoths and so on down to the smaller herd animals. Their die-back was a lot like what we see happening today to the wild animals of Africa at the hands of white and black "intelligent" beings. The meteorite seems to have been only one of the last straws for the great reptiles.

Might the demise of the dinosaurs, then, be part of a hidden pattern? Is it possible that an alien colony began a process of extinction that was by the meteorite (or meteorites) only finished?

A natural planetfall can't be distinguished from one targeted against ground settlements of a technological species. Is it possible that the dinosaurs were innocent bystanders in a genocidal war amongst alien settlers in the solar system?

The bombardment might only have been the last act in a more gradual ecological catastrophe that began half a million years before, when settlement of the planet resulted in extinction of species after species.

The introduction about this time, of flowering plants is another environmental perturbation that had profound ecological effects. It's not absurd to imagine this fitting into an overall pattern of outside intervention.

The settlement of Earth by a spacefaring race about seventy million years ago, then, offers one more (admittedly tenuous) explanation for the destruction of the higher terrestrial life forms over a brief period.

If we make this hypothesis, however, where are the traces of this earlier technological occupancy? Over sixty million years of oxidation will destroy many artifacts, but certainly some might survive.

Who can say? The cities we look for may lie beneath astroblemes. A look at a geological map of the Earth shows that continental plate boundaries have proved to be choice living sites. These plate-edge regions have suffered pronounced geological changes that could have erased most traces of alien settlement.

The final test of this hypothesis would be found among the planetoids of the solar system. The asteroids might hold remnants of visits to our star by extraterrestrials . . . perhaps whole cities, the leftovers of great populations: killed off, perhaps, by biological warfare in desperate retaliation by the Earthbound cousins they had annihilated.

CYCLES OF RECOVERY AND EXPANSION

This hypothetical explanation for the Cretaceous Mystery should merely take its place in a catalogue of possibilities, perhaps near the bottom. Still, it's interesting to note that the period since that catastrophe—an interval which culminated in the development of *Homo sapiens*—is the same sixty million years suggested by Jones and others for an optimum minimal galaxy-filling time by a technological race.

The Cretaceous-Tertiary event was not the only one of its kind. At least four other mass extinctions are found in the sedimentary record, including one at the end of the Devonian and another at the Permian-Triassic boundary, approximately 225 million years ago. These events are less well understood and may have taken place over longer periods than that of the Cretaceous, but we may compare the rough 10- to 500-million-year intervals seen with those suggested by Newman and Sagan for Galaxy filling by space-travelling species.

If the ecological holocaust of the Cretaceous was a local manifestation of the death spasm of a prior spacefaring race, whose overpopulated sphere of settlement spoiled and self-destructed as the shell of colonization passed outward, then we humans may have come into being almost too late. Any later, and the next wave —the expanding shell of still another spreading technological race—might have washed over Earth before we had the ability to assert property rights . . . assuming we have that ability now.

We may wonder if the Earth is the first Nursery World to have recovered sufficiently, since the last wave of "civilization" passed this way, to develop a species with intelligence. Whether or not the end of the Cretaceous corresponded to the agony of dying starfarers, it may well be that colonizing cultures inevitably leave behind them wastelands empty of intelligence and living voices.

If we humans initiate an era of interstellar travel of our own, we may find all around us the blasted remains of an earlier epoch. Would we then learn a lesson? Perhaps. But with the ever-present opportunities for expansion, those humans who exercise self-restraint and environmental sensitivity toward their adopted worlds will not be able to force this tradition upon those who travel far away to establish newer colonies. A nucleus of selfishness is likely to expand more rapidly than a center of more rational colonization. While there may be zones where settlers preserve and protect the local ecospheres, cognizant of their long-range potential, others may be rapacious.

Certainly our environmental record here on Earth is a test. The list of extinct species, some of which might one day have become starfarers, is long and growing longer.

The Great Silence may be the sound of sands drifting up against monuments. It may be quiet testament to the fate of species which allow "population pressure" to be their motivation for the stars.

MORE IDEAS

We'll begin a "morphological" analysis of the Great Silence by presenting the following list of possibilities:

1) Solitude.   We are unique in evolving technological intelligence.

This hypothesis implies something is very wrong with current use of the Drake Formula. Habitable planets may be rare, or some "spark" may be needed to initiate life out of a pre-biotic soup.

The final step to intelligence may require some software miracle that makes it far more improbable than currently thought.

Alternatively, the last term in the Drake equation—the average life span of technological species—may be on the order of decades. This might be due to some "inevitability" of self-destruction, or due to the "Deadly Probes" of Saberhagen and Benford.

2) "Magical" Technology. It may be that technological species soon discover techniques that make radio and even colonization irrelevant. We may be on the verge of such discoveries right now, though it's hard to imagine any race totally abandoning the electromagnetic spectrum, whatever its other options.

3) "Quarantine." The hypothesis of purposeful avoidance of contact.

This is an idea long popular in science fiction. It explains the Great Silence by suggesting that the solar system is kept as a "zoo." Or benevolent species might want to let Nursery Worlds lie fallow for long periods, to nurture new sentience.

Related ideas are that observers are awaiting mankind's social maturity or have quarantined us as dangerous, perhaps infected.

Kuiper and Morris also have suggested that members of a galactic radio club would not contact "beginners" because this would wreck our usefulness as members of the network. Making us information consumers too early would spoil us as information *providers*, whose unique experience would add richness to galactic culture.

ETIS may visit the Solar System for reasons having nothing to do with us.

A problem with "Quarantine" is the galaxy's differential rotation. Our neighbors don't remain our neighbors. If during one epoch we live near environmentalists, ten million years later our sun could enter the domain of a less scrupulous race. The Quarantine hypothesis appears to call for some degree of cultural uniformity in the galaxy . . . hard to accomplish in a relativistic universe.

4) Macrolife. The abandonment of planet-dwelling as a lifestyle.

Expansion will generally come from those colony worlds most recently settled. There might be a great selective process favoring those individuals suited to living in starships. One can imagine the pioneers eventually deciding that planet-bound existence is filthy and degrading. This might result in either of two different behaviors, each compatible with the Great Silence. Truly space-borne sophonts might greedily fragment terrestroid planets for building material and volatiles, leading to disastrous versions of "Solitude" or "Low Rent" (see below), or they might cherish Nursery Worlds for what they are and protect them, as in option "Quarantine," without any conflict of interest or desire to use high-gravity real estate.

5) "Seniors Only." More alternate lifestyles.

It's often suggested that spacefaring sophonts might "graduate" to other interests after a reasonable time. This would set a limit to the period of expansion, though not, perhaps, to exploration.

Discovery of immortality could tend to promote conservatism, and an aversion to the dangers of spaceflight.

6) "Low Rent." Earth is inaccessible or undesirable.

Spacefaring sophonts that otherwise had the means might choose to bypass Earth. A few possibilities to consider are the following.

a) The one technique for travel faster than light (FTL) which has drawn some support from the physics community has been "geometrodynamic"—via controlled entry into the zone of influence of a Black Hole and traversing spacetime through hyperdimensional shortcuts. If such a version of FTL travel were possible, convenient, and efficient, one might expect galactic civilization to cluster around entry and exit points. Long-range slowboat technology would languish.

The fact, then, that astronomers have observed no nearby black holes may be a manifestation of the so-called Anthropic Principle. If a "usable" black hole were closer, the Earth would have already been settled, an ecological holocaust would have ensued, and we would not exist to observe the black hole. Thus the fact that we are here is consistent with a failure to observe nearby black holes.

b) Another systematic effect that might make for periods of inaccessibility is the migration of the sun around the center of the galaxy. We are currently on our way out of a gas-and-dust-rich spiral arm. In a few million years the sun will be in an "open" area, where there are few bright, younger stars. Spiral arms are home to the dense interstellar hydrogen clouds. These are thought required to run Bussard ram-scoops, but today that particular type of vehicle is falling into some disrepute. Besides, the clouds might also be hazards to other forms of travel.

c) Earth life forms rely almost totally on the left-handed isomers of complex organic proteins and amino acids. This might not be the case elsewhere. Should "dextro-" life dominate everywhere else, we might find Earth systematically avoided because there would be nothing here for prospective settlers to eat!

These are just a few examples of an endless supply.

7) Migration Holocaust. This category has received the most attention in this article. Transient occupation of a Nursery World by a techno-culture might cause extinction of local higher life forms, delaying the local upsurge of intelligence and resulting in a neighborhood so depleted that we are the first to recover in the nearby area.

CONCLUSION

The quandary of the Great Silence gives the infant study of xenology its first traumatic struggle: between those who seek optimistic excuses for the apparent absence of sentient neighbors and those who enthusiastically accept the silence as evidence for humanity's isolation in an open frontier.

As humanity grows up, we're finding out just how complicated the universe can be. We've seen that "Galactic Empires" have implications far beyond anything considered even by the science fiction of the past. The universe has many more ways to be nasty, if it so chooses, than we had thought.

Opportunities do not, however, have to be taken up. While the author does accept that elder species will necessarily be wiser and more restrained than contemporary humanity, he does suggest, and hope, that such noble races DO crop up from time to time. If such a culture lived long, and retained much of the strength and vigor of youth, it might have taught a tradition of respect for the hidden potential of Life to all subsequent spacefaring species.

It might turn out that the Great Silence we're experiencing is like that of a child's nursery, wherein adults speak softly lest they disturb the infant's extravagant and colorful time of dreaming.

REFERENCES

On Interstellar Travel Technology

Bracewell, R.N., *Nature* (London) 186, 670 (1960).

Forward, Robert, "Interstellar Flight Systems," AIAA Paper No. 80–0823 (1980).

Martin, A.R., "Project Daedalus—Final Report of the British Interplanetary Society Starship Study." *BIS*. A.R. Martin, ed. (1978).

O'Neill, Gerard K., *Physics Today* 27, 32 (1976).

On Possible Dispersal of Intelligent Life

Ball, John A., *Icarus* 19, 347 (1973).

Billingham, John (ed.), *Life in the Universe*, MIT Press, 1981.

Cameron A.G.W. (ed.), *Interstellar Communications*, W.A. Benjamin Inc., New York (1963), (1970).

Hart, Michael, *Q.J.R.A.S.* 16, 128 (1975).

Jones, Eric, *Icarus* 46, 328 (1981).

Kuiper, T.B.H., M. Morris, *Science* 196, 616 (1977).

Newman, William I., C. Sagan, *Icarus* 46, 293 (1981).

Shklovskii, I.S., C. Sagan, *Intelligent Life in the Universe*, Holden Day (1966).

On the Cretaceous Catastrophe

Alvarez, L.W., W. Alvarez, F. Asaro, H.V. Michel, *Science* 208, 1095 (1980).

Dr. Robert A. Freitas, Jr.

ALIEN SEX

JUNE 1982

REPRODUCTION is one of the most important and characteristic activities of living things, and in most of the more highly evolved forms of life on Earth; it involves the cooperation of two sexes, each with a distinct role. If other worlds have their own forms of life, will the same be true of them? Well, maybe—but maybe only up to a point. Not only the roles of the sexes, but their number and even the very existence of sex itself, may be open to question. The universe, after all, has a lot of room for variety.

Dr. Robert A. Freitas, Jr., was born in Maine but has spent most of his life in California. He earned degrees in both physics and psychology from Harvey Mudd College, followed by a juris doctor from the University of Santa Clara. While at law school, he developed a long-lasting interest in extraterrestrial life. His first *Analog* article was "The Legal Rights of Extraterrestrials," and several others followed on other aspects of the problem. He has been active in several organizations dedicated to the exploration of space and in lobbying for governmental support of such exploration. ∎

OF all the important things life forms do, self-reproduction seems quite unique. Deprive an animal of its food or drink, draw off its blood, or cut away its skeleton, and it dies. But prevent an animal from reproducing and, usually, nothing happens. The species may eventually become extinct, but the individual organism lives out its life span. Reproduction of self is an important asset but is not absolutely essential for life—even on Earth.

This is true despite protests that self-replication is somehow the entire point of biological activity. The vast majority of social insects never engage in personal self-reproduction, yet these species are extremely successful. The anatomy of domesticated turkeys has been altered by breeding for plumpness, so that these animals can no longer mate in the natural way and must be artificially inseminated with human help. A number of higher Earth species such as the mule are quite sterile, yet do not become extinct.

Indeed, an intelligent extraterrestrial race might lack the capability of individual direct self-replication. We might imagine two closely allied nonsentient alien species among whom, when a successful interspecies mating occurs (or in a special way or in a special environment), sterile *but intelligent* "mule" offspring are the result of the union. Clearly, there is no bar to the rise of intelligence in such a situation—the hybrid's brain mass, neural complexity, or level of organization may be qualitatively greater than those of its non-sentient parents. Our intelligent but sterile race would maintain its numbers by corraling and manipulating the "dumb" mixed parental population, much as stockmen raise choice cattle and stablemen breed champion thoroughbreds.

It is entirely possible that some very complex extraterrestrial living creatures may have no need to reproduce themselves at all, either personally or at the species level. One class of such beings might be self-creating but non-replicating organisms, analogous to very advanced robots capable of making continual repairs and of upgrading their own mechanisms periodically. Other nonreproductive life forms might increase their numbers simply by physically expanding and then dividing into pieces of various sizes—biomass increases as easily by growing to larger volumes as by replicating a large number of small originals.

There could even exist a race which evolves by means of acquired characteristics. Such life forms would neither die nor reproduce, but would instead modify their parts to survive in a changing environment. Selection would act internally on their constitutions, rather than on a succession of descendent organisms. The closest analogies, according to Dr. P.H.A. Sneath, are terrestrial soils, which don't reproduce in the usual sense but are complexly organized systems nevertheless. Soils respond to environmental changes, arise where there is rock and wind to erode it, and are virtually immortal. If ever they tried to "compete" with their neighbors, such soil-like organisms would blend together with a total loss of individuality.

Finally, reproduction is not a prerequisite for sex. Two dissimilar growth systems could trade genetic information about their expansion patterns, then each continue growing in a slightly different way. This would be an example of "sexual growth" without replication. Of course, self-reproduction does have many advantages. Whole-body duplication allows rapid dispersion into new niches and produces abundant biological alternatives upon which natural selection may operate. It is a telling observation that most complex terrestrial creatures are capable of self-replication. Assuming Earth is a typically exotic planet, we should expect that many, though certainly not all, extraterrestrials will be reproducers.

IS SEX NECESSARY?

If reproduction is a useful convenience for a species, sex seems almost pure luxury. Certainly, there is no fundamental reason why evolution and diversity cannot thrive in its absence. There is no universal law prohibiting asexuality.

In fact, asexuals can be vastly *more* prolific in the short run. Microorganisms churn out literally billions of copies in the space of a few hours, relying almost exclusively on such simple techniques as binary fission and budding. No "opposite sex" is customarily required. While it is true that many sexual species are also quite fecund, as a general rule fewer offspring are produced than among the asexuals.

Furthermore, asexual reproduction is good economics from the personal point of view. An organism which copies itself without sex passes undiluted its entire genetic heritage to its young. Offspring are exact duplicates of the originals. A bisexual parent, on the other hand, normally contributes only *half* of its own genes towards the construction of an offspring. The other half must be donated by the second parent. From the standpoint of the selfish gene, sex entails a rather poor profit margin in comparison to no-sex.

Except . . .

A completely asexual species produces a population of virtual duplicates, save an occasional mutation. Since variation is the raw material of evolution, and the lack of sex decreases the breadth of this variation, such creatures are a distinct disadvantage when competing with their sexual brethren. New genetic combinations in asexual species can accumulate only by a sequence of fortuitous mutations in the same family lineage. Asexuals must "stand in line" to wait for a series of rare mutations. Change spreads only slowly through the gene pool.

Sex allows the accumulation of variation in parallel, rather than in series. In a sexual species many new genes can spread rapidly throughout the population because gene-jumbling produces a novel combination (possibly of several new genes at once) with each act of reproduction. Rare mutations become more widely distributed. So great are the advantages of sex that even many normally asexual organisms have occasional sexual encounters to beef up the waning gene pool. This is especially true in particularly harsh or rapidly changing environments.

For example, both the freshwater hydra and the aphid reproduce asexually for most of the year. As winter approaches, with hard times ahead, these animals switch over to sexual reproduction. This ensures genetic diversity when the colonies disband and disperse with the arrival of cold weather.

In the billion years or so since its invention, sex has proven remarkably successful—if we are to judge from the fossil record of life on this planet. Sexual species dominate the animal world, and the most widespread and important groups are all but exclusively sexual in their mode of reproduction. What of the creatures of other worlds? We don't know whether all alien species must have chromosomes, genes, or some other information-carrying molecules—perhaps some extraterrestrials reproduce by a process akin to xerography. But two things are clear: Variability is the key to biological complexity and survival, and sex reshuffles the biological data deck nonpareil.

HOW MANY SEXES?

Not all Earth creatures are bisexual. Terrestrial biology offers several examples of multisexual reproduction. One interesting case is the lowly paramecium, which has between five and ten sexes depending on how you count. These are distinct mating forms which arise at different times under definite conditions, and which can only mate in certain specific combinations. Another example is certain quadrisexual fungi, notably *Basidiomycetes*, in which there are four distinct sexual groupings. Among the higher animals, greylag geese display an evolved sociobiological "behavioral trisexuality." One goose "marries" and mates with two male ganders. Multisexuality is clearly a viable alternative.

Why, then, are the vast majority of terrestrial sexual life forms bisexual?

The answer seems to be that one sexual partner is just enough to properly shuffle the genetic deck. Each healthy individual has a reasonable chance of mating with a member of the opposite sex. Apparently, two are both necessary and sufficient. More than this may seriously impair the chances for species continuity. The more sexes required for successful reproduction, the more difficult it is to bring them all together properly at just the right time. The greater the number of links in the mating chain, the greater is the chance that the species may become vulnerable to certain predators or other environmental severities, thus jeopardizing the future of the entire race. And it is not clear how, say, three sexes could generate variability very much more effectively than two.

So while extraterrestrial multisexuality cannot be ruled out, requiring more than two sexes for reproductive activity seems an unnecessarily complicated solution to a problem elegantly resolved using only two. It's a safe bet that bisexuality is the overwhelmingly dominant mode of sexual reproduction among the alien life forms in our Galaxy.

THE BISEXUAL UNIVERSE

Assuming that most sexually reproducing ETs will have just two sexes, bisexuality does not necessarily demand the existence of distinct male and female forms. A case in point is the black mold *Rhizopus nigricans*, which displays an unusual form of reproduction known as "heterothallism." This species of fungus is bisexual, inasmuch as two organisms are required for fertilization and replication to take place. However, the two sexes are physically indistinguishable. There are no constant differences between members of opposite mating groups other than their reciprocal behavior when crossed. Thus, it is impossible to designate one form of the black mold as male and the other as female. Customarily, the complementary groups are labeled merely "+" and "−" for convenience during experiments.

One can imagine a race of intelligent extraterrestrials apparently unisexual to our undiscerning eyes but which actually practice heterothallic sex. Such beings would most certainly lack secondary sexual characteristics, those hormone-induced physical landmarks such as beards and breasts to which we humans are so pleasantly accustomed. They might even lack distinctive primary sexual characteristics such as internal or external gonads. Norms of marriage, inheritance, language, religion, and social behavior would be profoundly affected by this state of affairs. The usual social tensions caused by sexual competition in human cultures would be more diffuse in a society in which every member was a potential mate and in which all could become pregnant, though sexual undercurrents might arise in all interpersonal relationships. The disparate male/female roles in human social roles and courtship rituals would defy their understanding, and to heterothallic ETs, human males—who participate in reproductive acts for pleasure but cannot become pregnant as a consequence—might be judged especially pitiful, handicapped, even perverted creatures.

Assuming maleness and femaleness exist among most bisexual alien species, there are again major variations in Earthly biology. It is quite possible to have an organism which is neither strictly female nor strictly male, but rather exhibits some alternating or intermediate condition. For example, simultaneous hermaphrodites possess at once both female and male sex organs. Ovaries and testes are present together in the same individual. Matings occur in pairs, with each partner serving both sexual roles at the same time. Planarians, earthworms, sponges, and snails fall into this category, and a few simultaneous hermaphrodites among the more highly evolved vertebrates are known, such as the banded flamefish *Serranus subligarius*.

Such intersexual animals can be sex-mosaics in time as well. Many creatures start life as one sex and finish it as another. These sequential hermaphrodites come in many varieties. For instance, in protoandry an animal is first male and later female; proterogyny is the converse, with young females metamorphosing into functional males as they age. Or the process can be cyclical. Oysters are born as males, then spend the rest of their lives switching back and forth between male and female in irregular cycles a few months long.

What would a society of sequential hermaphroditic aliens be like? We can take a few clues from the life history of the freshwater shrimp *Gammarus pulex*. Each of these individual crustaceans is both male and female, but not at the same time. Newborn animals spend early life in a neuter stage, after which they pass through puberty and enter the first sexually active phase as functioning males. After a while, the maleness is exhausted. Latent ovaries ripen into maturity, and the organism spends the remainder of its life as a full-fledged female. Eggs are shed by middle-aged mothers and are fertilized by energetic youthful males still in the middle of their first cycle.

It is a magnificent bisexual system, one which works quite well on Earth. No individual is excluded from any phase of the reproductive process. Still more significant, each member of the colony plays both male and female roles during his/her life. Drawing an analogy to the human life cycle, zoologist Norman J. Berrill of McGill University in Montreal imagines that all halfgrown individuals, about ten years old and weighing about 34 kilograms, would be males—the only males—ready to act as such both sexually and "probably in other wayward ways." Like their truly human counterparts, as troublemakers they would be kept in line by a closed society of matriarchs, roughly equal in number to the males but each twice the size and much older and wiser. This wisdom would be not merely of a general character, as among human parents, but also in the special sense of each having been a male herself, as understanding as a mother with a child and as little likely to put up with any nonsense, perhaps wistfully looking back to her youthful manhood. Womanhood would bud as usual when masculinity had faded, with growth continuing and full female maturity yet to come.

The institution of monogamous marriage as we know it would be quite impossible in such a society. Husbands would be forever changing into wives and males would be too immature psychologically to be treated as other than "child-lovers." Such pedophilia is viewed as a sexual perversion in many human societies, but for our intelligent shrimps it would seem quite normal. Incest prohibitions might be inordinately complex, since all fertile middle-aged females in the family, in theory, could mate with any or all male children. To offset the negative effects of inbreeding, exchanges of matriarchs could occur between families, doubtless accompanied by the same pomp and ceremony as upon "giving the bride away" in our society. Love in the traditional human sense probably would not exist—females could have strong affective and familial non-sexual ties with other females, whereas relations between females and males would be characterized more as controlling playfulness than by affectionate cooperation. Our usual concepts of male/female love might seem quite alien to them.

XENOGAMY

Given these tremendous potential cultural and biological differences, one wonders if meaningful interspecies social-sexual relations would be possible at all between humans and extraterrestrials. Many science fiction authors have tried to deal sensibly with this touchy question, such as Philip José Farmer in *The Lovers*, in *Flesh*, and in *Strange Relations*; Walter Tevis in *The Man Who Fell to Earth*; and a host of others. There have been "reports" of sexual molestations of humans by the occupants of UFOs. And "Star Trek"'s own Mr. Spock is a prime example of xenogamy, the product of a marriage between a human female and a male alien from the fictional planet Vulcan.

It is not at all implausible that interspecies copulation can occur. Given the prevalence of the complementary male and female organs throughout the animal kingdom on this planet, such activity may indeed be possible even between creatures of "gross morphologic disparity." Alfred Kinsey's researchers turned up accounts of attempted couplings between a female eland and an ostrich, a male dog and a chicken, a female chimpanzee and a tomcat, and a stallion and a human female. Obviously, relations between humans and other beings even roughly humanoid in shape can happen.

If such activity is possible, is it likely? Could humankind and an alien race derive sexual pleasure from mutual physical encounters? These are very difficult questions, mainly because the ET is such an unknown quantity. Extraterrestrials may have organs, appearances, sensitivities, and responses wholly incompatible with any conceivable human style of lovemaking.

And yet—in 1948 Kinsey reported that some 17% of all rural farmboys had experienced sexual congress with various barnyard animals, and had achieved orgasmic satisfaction in this way. (Less than a tenth of a percent of all females interviewed admitted such coition, although 1.5% of the sample reported *some* form of sexual contact with animals.) If bestiality occurs so regularly among human populations, can we state with any assurance that "xeniality" will not also occur when humans mingle socially with alien races? The evidence, scanty though it may be, suggests that interspecies sexual contacts are not only possible but probable.

One last question remains. When humans and aliens sexually join, will anything result from the union? Again, this is a difficult question because an unknown alien physiology is involved. Different species on Earth have been mated successfully from time to time—for instance, the hybrid offspring of a mallard and a pintail duck is fertile.

In 1975 a chance mating of two very different species of ape in the Grant Park Zoo produced the first reported ape hybrid. The offspring, dubbed a "siabon," was the result of a mating between a male gibbon and a female siamang confined in a single cage. "Obviously," remarked one researcher, "they had been sexually involved for some time." Gibbon cells have 44 chromosomes, whereas siamang cells have 50, and thus the two are farther apart genetically than human beings and the great apes. The "siabon" offspring, believed sterile, has a mixed bag of 47 chromosomes—22 from the father and 25 from the mother. Still, in the first analysis, xenobiologists recognize that interspecies fertilization, and especially hybrid fertility, is a rather rare phenomenon.

In the context of extraterrestrial matings, natural interspecies fertility should be even rarer. (Of course, with advanced technology almost anything may be possible—the first interkingdom clones combining plants and animals were achieved during the late 1970s.) We know that slight changes in the environment can cause enormous variations in planetary biochemistry. Nucleic acids, genes, and codons may not be needed by ETs, or these may be essential but in different forms than are found on Earth. Many complicated and highly unlikely coincidences must occur for an alien/human mating to produce viable results. The two species must have identical amino acid sequences for proteins (assuming they even have proteins), the same optical rotation in their biomolecules, closely matched chromosomes with similar size and shape, the same kinds of genes located on the same chromosomes at the same locations, and so forth—all of which is highly improbable. It has not even been shown that humans can produce interspecies offspring with their own closest biological relatives—apes and other primates who share most of man's biological heritage.

So interspecies matings involving humans aren't likely to result in pregnancy. If pregnancy somehow does occur, the hybrid offspring probably won't be viable. (It has been estimated that up to 50 percent of all *normal* human pregnancies may end in spontaneous abortion.) Finally, if somehow viable and carried to term, the interspecies hybrid will most likely be sterile or maladapted for natural survival, much like the mule or the liger. Hybrid vigor is unlikely in the offspring of parents of such widely varying genetic constitution.

This does not augur well for Mr. Spock.

REFERENCES

Berrill, Norman J., *Worlds Without End: A Reflection on Planets, Life, and Time*, The Macmillan Company, New York, 1964.

Freitas, Robert A., Jr., "Xenobiology," *Analog* 101(30 March 1981):30–41.

Kinsey, Alfred C., et al., *Sexual Behavior in the Human Female*, W. B. Saunders Company, Philadelphia, 1953.

Smith, J. Maynard, *The Evolution of Sex*, Cambridge University Press, Cambridge, 1978.

Sneath, P.H.A., *Planets and Life*, Thames and Hudson, London, 1970.

Part Five

COMING

SOON...

Arthur C. Clarke

NEW COMMUNICATIONS TECHNOLOGIES AND THE DEVELOPING WORLD

MARCH 1982

WE in North America and Europe are aware of the changes such innovations as telephones, radio, computers, and communication satellites have made in our lives (though it's surprising how quickly we come to take them for granted and then to forget how different things were before). What impact, if any, will these advanced technologies have on the Third World, where survival itself is the foremost problem on most people's minds? The answers may surprise you.

The article that follows is based on an address the author gave at a conference of the International Program for the Development of Communications at UNESCO headquarters in Paris. Arthur C. Clarke is uniquely qualified to speak and write on this subject. Probably best known in this country as one of the most popular of all science fiction writers, he was educated in physics and has had a long-standing, active interest in communications. (In 1945, he developed the basic theory of communication satellites—long before the world was ready to do anything with the idea.) He has also lived for 30 years in a developing country, Sri Lanka, so he has firsthand familiarity with the problems such countries face. He is chancellor of the University of Moratuwa, which is also the location of the Arthur Clarke Centre for Modern Technologies established by the Sri Lankan government. ■

IN many ways, and for many purposes, printed matter—books, newspapers, wallposters—will always be the best and cheapest form of communication. But now electronics has given us tools that can perform miracles impossible to the printed word—and which, of course, can reach millions who are unable to read. The newest and most powerful of these communications devices depend upon space technology, and because this fact is not yet generally realized, I shall be concentrating upon it here.

But first, a few basic considerations: What does a country need in the way of electronic communications? At the risk of stating the obvious—which is often not a bad idea anyway—I will list them in order of priority.

1. THE TELEPHONE

A reliable telephone system must surely have the first priority: it affects every aspect of life—personal, business, government. We have also belatedly realized that it can be a major energy-saver, making countless journeys unnecessary. It is also the greatest *life-saver* ever invented, though often it requires a tragedy to bring that point home. A few years ago, thousands died when a dam burst in an Asian country. The telephone that should have warned them was out of order. . . .

It will be a long time—though not as long as you may think—before everybody has a telephone. But with a telephone in every village, we can have the next best thing. Telegrams can be dictated so that anyone can get a message to anyone else within a few hours, with all the implications this has for business and social life.

2. RADIO

There is no need to stress the value of radio, both for spreading information and establishing a national consciousness. If I were dictator of a country so poor that it had to choose between a telephone system and a radio service, I would be tempted to put my money on the radio—despite everything I've just said about the importance of telephones.

And radio is nowhere near the end of its development. We have seen the transistor revolution of the 1960s merge into the solid-state revolution of the 1970s, so that radios are cheaper and smaller and more reliable than anyone could have dreamed, even thirty years ago. I can see at least two more revolutions ahead. First there will be built-in solar cells to generate electricity, so that we won't have to bother about batteries any more—what a boon that will be in remote places! Second will be the coming of direct-broadcast satellites, so that perfect reception will be possible over all the world, and the horrid cracklings of the short waves will be a thing of the past.

3. TELEVISION

Everything that can be said about radio is true of television, squared, so most of what I say will concentrate on this medium.

4. TELEX

Telex equipment is still very expensive and limited to commercial and government use. However, recently developments in the home computer field have shown that it could be quite cheap. It will eventually merge into—

5. DATA AND COMPUTER NETWORKS

Even for highly developed countries, these are still in their infancy, though they undoubtedly represent the wave of the future. It will be a long time before we in the Third World can afford them; of greater importance to us may be what I have named—

6. ELECTRONIC TUTORS

These will not involve communications links at all; they will be the next generation's equivalent of today's pocket calculators, and will be about the same size and cost. They could trigger as big a quantum jump in mass education as did the invention of printing five centuries ago.

But let me start by going back to 1965, when the first commercial communications satellite, Early Bird (INTELSAT 1), went into orbit. It could carry 240 telephone circuits, or a single television channel.

Fifteen years later, in 1980, INTELSAT V was launched. It carried not 240, but 12,000 simultaneous phone calls—at a fraction of the cost. *And* several TV channels at the same time.

However, we are now reaching the limits of what can be done by purely robot satellites. Comsats as large as tennis courts can just be squeezed into existing launch vehicles, to unfold like glittering metal flowers when they reach space. But in another decade, we shall need satellites as big as football fields—ultimately as large as cities (indeed, some of them will be cities!). They will become possible thanks to manned transportation systems like the space shuttle, which can carry construction crews and their equipment into orbit.

But why do we need such huge satellites—what have they got to do with the problems of the Third World? The answer may seem paradoxical, even perverse.

In highly developed regions like the United States and much of Europe, communications satellites are a great convenience, but are not absolutely vital. These countries already have excellent cable and microwave links.

To many developing countries, however, satellites are *essential*; they will make it unnecessary to build the elaborate and expensive ground systems required in the past. Indeed, to such countries, satellites could be a matter of life and death. To put it as dramatically as possible, unless major investments are made in space, millions are going to die, or eke out brief and miserable lives. And most of those millions will be in the Third World.

Let me explain this paradox, which is typical of the way in which technology affects modern society—and is why no one without *some* understanding of these matters should be allowed to enter the corridors of power.

Because the first comsats were small and feeble, it was necessary to build huge, multi-million-dollar ground stations, with dishes thirty meters across, to contact them. Thus their sole use was to provide links between national telephone, telex, and TV networks—where these existed. They transformed the pattern of world communications, but did not directly affect the man in the street—still less the man in the mud hut.

That situation is changing with explosive speed. When only a few score Earth stations were involved, it made sense—indeed, there was no alternative in the 1960s and 1970s—to put the complexity and expense on the ground. But now that there are larger and more powerful satellites in orbit, ground stations can be much smaller and cheaper. Indeed, for the simplest ones the cost has been reduced a *thousandfold*! All over the United States there are now homes with dishes about three meters across, picking up scores of programs from the communications satellites hovering high in the southern sky. Soon these dishes will be less than a meter across, and everyone who can afford TV at all will have them. This is the beginning of the DBS—*Direct Broadcast Satellite*—revolution. It means that ultimately a few very large satellites can provide any type of service—telephone, telex, television, data, computing facilities—at extremely low *per capita* cost to every member of the human race . . . except for those rather few people who live near the North or South Poles. . . .

I'm not saying that satellites will do everything. In heavily populated areas, fiber optics and short-range radio or infrared broadcasts will often be preferable. But even these local systems will, of course, be linked to the global network through

satellites. And *only* satellites can provide every conceivable type of communication cheaply and efficiently over entire continents, and to all moving vehicles on land, sea or in the air.

One of the first persons to realize the implications of this for developing nations was the late Dr. Vikram Sarabhai, whom I first met at the UN Conference on the Peaceful Uses of Outer Space at Vienna in 1968. In the paper he delivered there, he stressed the importance of "leapfrogging" obsolescent technologies and—in some cases at least—going straight to advanced ones. We have seen this happen in transportation; many nations have bypassed the railroad age, and gone directly from ox-carts to aeroplanes.

As you may know, Dr. Sarabhai was the driving force behind the first large-scale attempt to use a communications satellite—the NASA/Fairchild ATS 6—for direct TV broadcasting to rural areas. Unfortunately, he did not live to see even the initiation of that daring experiment, which took place in 1975–76. It was brilliantly carried out by his colleagues of the Indian Space Research Organization and the Space Applications Center at Ahmedabad, headed by Dr. Yash Pal—who, I am happy to see, will be Secretary General of the next UN Space Conference in 1982.

Some of the remarks that Dr. Sarabhai made at the '68 conference deserve to be repeated now; the intervening years have made them even more timely:

> *A developing nation following a step-by-step approach towards progress is landed with units of small size, which do not permit the economic deployment of new technologies. Through undertaking ventures of uneconomic size with obsolete technologies, the race with advanced nations is lost before it is started. . . . Developing nations such as India have the possibility of effectively using space communications for national needs. Compared to advanced nations . . . they have indeed an advantage* through not having an existing major investment in older technologies (my emphasis).

Of course there are problems, and to continue quoting Dr. Sarabhai:

> *We often meet with a lack of self-confidence to pursue major tasks involving complex and unfamiliar technologies . . . anything which is innovative . . . is automatically regarded with suspicion. The administrative structure of governments in many nations is dominated at the top not by technocrats but by professional administrators, lawyers, or soldiers, who are hardly likely to provide the insights, experience, and the first-hand knowledge of science and technology which are necessary at the decision-making level. Moreover, advanced nations often play a negative role in their interaction with the developing countries. There is seduction by their political and commercial salesmen who dangle new gimmicks which they suggest should be imported . . .*

That last warning of Dr. Sarabhai's is one that I would like to endorse—and amplify. No technology in the history of mankind has ever produced so many hypnotically irresistible gadgets as the electronics business. Radio telephones, visual display units, talking calculators, video recorders—the list is endless. And six months after one device comes on the market, it's superseded by a model twice as good at half the cost.

Thus any developing country wishing to take advantage of all these marvellous new facilities must proceed with great caution. There is, as Dr. Sarabhai warned, always grave danger of becoming locked into an obsolete technology. You must be particularly wary when something is offered you for free—there may be a catch. Even with the best will in the world, and the most expert advice, it's sometimes impossible to avoid bad decisions. The speed of technological development is so swift that the better is the enemy of the good, and the best is the enemy of both.

At this very moment, in the vital field of video recording, such a complex multi-billion-dollar battle of standards is raging that a wise buyer will wait to do business with the survivors. We now have three main TV standards—PAL, SECAM, and NTSC. There are three competing videotape systems—Betamax, VHS, and Phillips—with rumors of a much more compact one (longitudinal recording) on the horizon. Also, two incompatible video *disc* systems are about to come on the market. And in the United States, CBS has just demonstrated high-definition TV using over a thousand lines. . . . If you're not thoroughly confused, you've not understood what I'm saying.

Fortunately, the development of global networks will eventually compel some degree of standardization, and many of the improved systems will still be compatible with the old ones, as black-and-white TV sets can still work with color transmissions. Nevertheless, I don't envy anyone who has to advise his country what to buy—or to accept as a gift—in the telecommunications field during the next few years, or, for that matter, for the rest of the century. By 2001 *everything* we have now will still be operating somewhere. And it will *all* be obsolescent.

Let us ignore these messy practical problems, which can and must be solved, and look at the wider view. We are now entering an era when any conceivable type of communication or information could be available to any individual, anywhere on Earth, at any time. The only constraints are economic and political, not technical.

Like all technologies, the ability to communicate is neutral; it can be used for good or for bad. There was an amusing example of this recently, not a thousand kilometers from Sri Lanka, when the venerable profession of piracy had a sudden revival. A small ship was forced to anchor, owing to engine trouble, at a remote Indian Ocean island—and a flotilla of local canoes promptly descended upon it. In no time, everything that wasn't screwed down—and many things that were—had been spirited away, despite the anguished protests of the crew.

How did these latter-day pirates—these unsophisticated islanders—learn so quickly about their windfall? Simple; they'd all called each other up on their Citizen's Band radios. . . .

Of course, this doesn't mean that CB radios should be prohibited any more than one should ban telephones because countless crimes have been committed with their aid. We must accept the good with the bad, unless we assume that the invention of speech was a big mistake in the first place.

At this point I would like to draw a distinction between *regulation* and *control* in the field of communications. Regulation of some sort is inevitable and necessary for the good of all users, just as one has to decide whether to drive on the right or the left side of the road. Regulation is essential to ensure standardization and because spectrum space is limited; the International Telecommunications Union has been doing this task with considerable success for over a hundred years.

But *control*—i.e., the management of the *messages*, not the *medium*—that is another matter. It ranges all the way from complete censorship to well-intentioned cultural guidance. Let me quote my friend Dr. Yash Pal on this:

> In the drawing rooms of large cities you meet many people who are concerned about the damage one is going to cause to the integrity of rural India by exposing her to the world outside. After they have lectured you about the dangers of corrupting this innocent, beautiful mass of humanity, they usually turn around and ask: 'Well, now that we have a satellite, when are we going to see some American programs?' Of course, they themselves are immune to cultural domination or foreign influences.

I'm afraid that cocktail-party intellectuals are the same everywhere. Because we frequently suffer from the scourge of information pollution, we find it hard to imagine its even deadlier opposite—information starvation. I get very annoyed when I hear arguments—usually from those who have been educated beyond their intelligence—about the virtues of keeping happy, backward peoples in ignorance. Such an attitude seems like that of a fat man preaching the benefits of fasting to a starving beggar. And I'm not impressed by the attack on television because of the truly dreadful programs it often carries. Every TV program has *some* educational content; the cathode ray tube is a window on the world—indeed, on many worlds. Often it's a very murky window, but I've slowly come to the conclusion that, on balance, even bad TV is preferable to no TV at all.

In this connection, let me quote a testimonial from an unexpected source. During the late 1950s, South Africa was the only wealthy country in the world which did *not* have a television service. The minister in charge of broadcasting adamantly refused to permit one. "Television," he proclaimed, "will mean the end of the white man in Africa."

That was an extremely perceptive remark. From his point of view, the minister was perfectly right. If the pen is mightier than the sword, the camera can be mightier than both.

No wonder that *all* governments, whether they are liberal or not, make some attempt to control what appears on television. Indeed, there is material which virtually everyone would agree should be kept out. Sadistic pornography and incitement to violence against racial or religious minorities are obvious examples.

Back in 1960—five years before INTELSAT 1—I published a short story about a plot to brainwash the United States with the help of communications satellites broadcasting pornographic programs. I wrote it with the deliberate intention of making people think about the potential of comsats, both for good and for evil.

There is a vast territory where even men of good will may disagree fundamentally on what should, or should not, be presented to the public. Exposures of scandals or political abuses—especially by visiting television teams who go home and make rude documentaries—can be painful, but also very valuable. Many rulers might still be in power—or even alive—had they known what was really happening in their own country. A wise statesman once said: "A free press can give you hell, but it can save your skin." That is even more true of TV reporting—which, thanks to satellites, will soon be transformed out of all recognition.

Last month I had the pleasure of showing my old friend Walter Cronkite around Sri Lanka, while we filmed one of his "Universe" programs. I say "filmed," but actually we were using electronic cameras, and it was wonderful to view what we had shot within minutes instead of days.

However, even the electronic cameraman still has to get his cassettes through an obstacle course of postal authorities and customs officials and censors. But not for much longer; very soon he will need only a small collapsible dish, about the shape and size of a beach umbrella, and he'll be able to beam his pictures up to the nearest satellite, and straight to home. . . .

The implications of this are enormous. Just one example: how many soldiers would shoot a cameraman if they knew that millions of people were watching? And if you think some countries would not admit TV teams under these conditions—well, as equipment becomes so compact that a single man can carry it, the more difficult it will be to keep him out. And the harder closed societies try, the harder they will have to explain what it is they are so anxious to hide. In the end, they'll give up.

You doubt this? Then let me remind you of one astonishing step which has already occurred in this direction. Could you imagine, twenty years ago, someone in the Pentagon asking his Russian counterpart: "Would you mind if we photograph the Soviet Union from end to end, at such resolution that we can see everything bigger than a football?"

Remember the uproar, back in 1960, when Gary Powers's Lockheed U2 was shot down over Russia, doing precisely this! Yet now it is happening every day, without a murmur of protest from either side. Reconnaissance satellites are of such benefit to both parties that they are accepted by mutual consent.

In the same way, it will have to be recognized that all types of information and communication (with some obvious exceptions) are of benefit to everyone. Truth will out eventually, and those who try to suppress it will be condemned by history, as a recent United States president discovered to his cost.

In the struggle for freedom of information, technology, not politics, will be the ultimate decider. One turning point came several years ago, though few people realized it at the time. With the introduction of IDD (International Direct Dialing) the power of the state to control news was irrevocably broken. Private individuals could speak to each other across frontiers, and though Big Brother might catch some of them, he couldn't possibly keep track of them all.

But this is just a beginning, for the age of the telephone as a fixed instrument is swiftly passing. One of the main objectives of the very large comsats now being discussed is the provision of mobile, person-to-person communications. The old science-fiction dream of the wristwatch telephone could soon come true, and at a cost of a few dollars—given the determination to achieve it. Can you imagine the impact of this upon societies where, at present, there is only one telephone per thousand people? Some time during the next century, the human race will become one big, gossiping family. . . .

As the world comes to depend more and more upon satellites not only for all types of communications, but weather forecasting and resources inventories, search and rescue, navigation, etc., etc., the developing countries will be faced with a problem clearly foreseen by Dr. Sarabhai. I quote again from his speech to the 1968 Vienna Conference:

> *One of the hardest questions to be faced in adopting a satellite for national needs arises from the fact that many interested nations would not expect in the near future to have an independent capability for placing such a satellite in orbit. . . . The political implications of a national system depending on foreign agencies for launching a satellite are complex. . . . Perhaps collaborative participation of nations in the construction and operation of a launching system for the peaceful uses of space would be realized in the long run. The military overtones of a launcher development program of course complicate the free transmittal of technology . . . but knowledge cannot for long be contained within artificial boundaries and one has to learn to share. . . . Restrictions on the transfer of technologies which are involved in the peaceful uses of outer space merely jeopardize the security of the world through retarding the progress of nations.*

Any Third-World nation wishing to have its satellite launched will soon have quite a number of options. Currently, the most reliable haulage firms are the US and the USSR, either of whom will be happy to quote prices . . . at least to their friends. Many years ago the president of COMSAT remarked to me in his Washington office: "If the Russians offer me a better deal than NASA, I'll accept." Before long the European Space Agency, China, India, and Japan will also be in the market. And there are a couple of dark horses trying to enter the race with cut-price launchers—Lutz Kayser's controversial OTRAG operation and Gary Hudson's California-based company.

So five years from now there will be no lack of vehicles. I would suggest that a choice be made on purely pragmatic grounds—what insurance premium does Lloyd's quote on the launch? I am not joking—payload insurance is now very big business.

Of course, most developing countries will be concerned neither with building nor launching satellites, but merely renting facilities in them—as they do now. There will be more and more specialized satellites shared by countries in the same geographic region, even countries which, down on Earth, are not very friendly with each other. Radio waves have never respected frontiers, and from an altitude of 36,000 kilometers, national boundaries are singularly inconspicuous. The world of the future will be an open world.

I have space now for only a brief reference to another electronic development which is just about to burst upon us, and which may be at least as important to the Third World as communications satellites.

Little more than ten years ago, pocket calculators came on the market. Instantly, slide-rules and mathematical tables became obsolete; engineers, scientists, businessmen—*everyone* who had to work with numbers had his life transformed. The first calculator cost several hundred dollars; now you can buy far superior models for twenty.

Today, a second and even more momentous revolution is just starting based on the same technology. This is the advent of the electronic *book*—a whole encyclopaedia—even a whole library—in the palm of your hand. The pocket translators already on the market just hint at the possibilities; these electronic books will also speak, so that they can, for example, teach a foreign language. Beyond that, they will have plugged-in programs, so that they can provide tutoring in virtually any subject. They will be nothing less than electronic educators—able to work twenty-four hours a day. Of course, no machine can replace a good *human* teacher—but no country ever has enough of those! When they are made in the millions, the electronic tutors will cost no more than the pocket calculators of today. Solar-powered, they will need no batteries; properly designed, they will never wear out. So their written-down cost will be negligible, and even the poorest countries will be able to afford them, especially when the reforms and improved productivity they stimulate help those countries to boot-strap themselves out of poverty.

To sum up—what I have tried to do is to sketch some of the truly astonishing communications possibilities of the next decade. Almost everything I have described will be commonplace by the year 1990—all of it will be available by 2000.

I am well aware that many of these electronic marvels may serve only to increase the sense of frustration in countries where schools can't even afford blackboards and chalk. But more and more we must think of the human race as a single unit—and Mankind can afford anything it wants, especially if it stops squandering its resources on weapons of destruction.

What I have described to you are the weapons of peace.

Mark E. Peeples, Ph.D.

HUNTINGTON'S HANDLE

OCTOBER 1987

THERE are several diseases, such as Huntington's chorea, which we cannot cure, at least so far, but which we may soon be able to predict. They are transmitted genetically in such a way that a person who receives certain genes has a high probability of developing the disease at some point in life—and transmitting the same susceptibility to his or her children, if any. At present, people from families in which these genes are found must make important decisions based on educated guesses. Soon, as described here, they may may be able to make those decisions on the basis of definite knowledge—but will they want that knowledge?

Mark Peeples, with a B.S. from Heidelberg College and a Ph.D. from Wayne State University Medical School, did postdoctoral research on viruses at the University of Massachusetts Medical School in Worcester. He is now an associate professor of immunology and microbiology at Rush-Presbyterian-St. Luke's Medical Center in Chicago. This article was inspired in part by personal experience, as he and his wife saw the story of "Lou and Judy" lived by two of their best friends. The names and some of the details have been changed, but the essence of the story is true. ∎

I WAS the oldest. I was closest to my sister Judy, two years younger than I. I guess I always felt protective of Judy, which I didn't have to since she was so damn bright and independent. I wanted the best for her. I goaded her into doing her best.

When Judy was finishing high school, I was in my second year of college. As always, I was wiser for my mistakes and successes and wanted to share my insights with her. I encouraged her to go to the best school with the broadest opportunities in our home state, Ohio. I had decided that that school was Miami University. I had a friend at Miami, so, Judy and I visited the campus for several days in the early spring. Judy loved it and immediately applied for admission and was accepted. My favorite person had a chance to develop her full potential, which I knew would be great. Now I look back on that "help" and wonder if things might be different if I had not meddled in her life.

Judy did great in school, as I knew she would. She decided on a course of elementary education. She had always loved kids. She saw herself teaching first-graders and writing children's books someday. I thought that she would follow a more specialized, intellectual line, perhaps literature or psychology. Her response was thoughtful: love of art and literature, or stability and caring develops early in a child and needs encouragement. When I watched her deal with our cousin's little monsters at Christmas, I could see that she loved them and was a natural teacher, and, someday, mother.

Judy wanted a big family; she wasn't sure if six kids was enough. As a member of "Zero Population Growth," I was horrified. But again, her gift with kids was so obvious that I was willing to admit that six great kids might be an asset to our world. Of course, one major obstacle remained between Judy and her dream of six kids—she needed a husband. But there was time. And after all, that's what college is for.

As our family reunited for the annual Fourth of July picnic, in hometown Shelby, I met her "new" friend, Lou. Actually, Judy had known Lou since her first year at college and they had been dating for more than a year. During the few days of the visit, I grew to tolerate this 'thief" and actually to like him. He was interested in my science, and we had a common interest in Rolling Rock, brewed one state to the east of us. I saw Judy and Lou several times during the next year. Each time, I saw them becoming closer. I enjoyed their company. I began to see them as a natural couple.

It was really no surprise when Judy and Lou announced their engagement at their graduation party. The wedding was to be a year later. Both had found jobs in the Columbus area. The year between graduation and their wedding would give Judy a chance to start her career teaching first-graders and Lou a chance to work with a prestigious architectural firm in Columbus. Both needed the time to establish themselves, and the savings to start a life together.

Late one February night, Judy called me. Lou had left. Disappeared. No forwarding address. She had called his father, but was told only that he was gone. Okay, but gone where? No answer; Lou wanted it that way. Judy became more and more desperate as she spoke. What had she done? We talked for two and a half hours without obvious answers. Nothing seemed wrong. Judy had been so happy with her job of creating people out of first-graders that she and Lou had been discussing how many and how frequently they should have kids. That dream had vanished.

During the next several months I talked to Judy by phone nearly once a week. The puzzle continued, the guilt feelings persisted. By their proposed wedding date in June, Judy had given up hope of ever seeing Lou again. It was difficult for her to think of another man. She had been so certain. . . . But she had to move on and began dating again. Within three months, she was steadily dating Joe, who taught phys. ed. in the same school system. They seemed compatible enough, but he was no Lou. Judy told me over and over that she was comfortable with Joe and that she was anxious to start her family. They were engaged.

Lou reappeared, as suddenly as he had disappeared. At first Judy would have nothing to do with him, but finally, Lou convinced her that she should listen to his explanation. His story sounded unbelievable, but it was true.

Lou's mother had died when he was 16. His first memories of her were happy enough. But he slowly began to realize that she was "sick." That was his Dad's word to explain the dishes she dropped, her forgetfulness, or her sudden furious outbursts. Then the dancing started, the sudden involuntary twists and jerks. She frightened Lou. She told Lou that she felt like a puppet with an unseen demon at the controls. He could barely understand her speech. Her eyes would roll, her eyebrows jump up and down, and her tongue bounce in and out. She spent more and more time in deep depressions. The family had watched, and cared for her, as she fell from an active, cheerful provider to a noncommunicating grotesque vegetable. Lou's mother died of pneumonia at the age of 45, after inhaling some food.

The family doctor and a neurologist had diagnosed her condition as Huntington's disease, also known as Huntington's chorea, from the same root as *choreograph* because of the dance-like movements late in the disease. Because folksinger and songwriter Woody Guthrie died of this disease (not of schizophrenia as diagnosed by his doctors), Huntington's disease is also called Woody Guthrie's disease. Twenty thousand Americans suffer from Huntington's disease. A hundred thousand are at risk.

Lou had experienced, first-hand, the terribly frustrating and painfully slow demise of his mother. He also knew that his mother's father had met with the same fate. When Lou's mother died, the doctor explained to him that Huntington's disease was hereditary, a dominant trait. Careful scrutiny of his family tree showed several other instances of apparently similar degenerative disease.

Lou knew that "dominant" meant that he has a 50% chance of carrying the same Huntington's disease gene. If he has inherited the disease gene, his disease will be similar to his mother's. Did he inherit the gene? There was no way to know. He is now 26. The disease could begin anytime in the next 30 years. Everytime Lou drops a glass or trips over a step, he thinks that it may be starting. The pressure is tremendous on Lou for his own life. But what about the strain he would put on Judy, if they were to marry? He knows how difficult it was for his father to cope with his mother's long illness and he doesn't want to put Judy through that.

Judy was stunned. Lou had never talked much about his mother. He had only said that she had become ill and died when he was in high school. It was obviously painful, so Judy had avoided the subject.

Judy realized that Lou hadn't deserted her for anything she had done or for any selfish reason. He had needed time to think and decide what was most important in his life, and in Judy's life. Now that Lou was back, Judy also realized that she still loved him, perhaps more than before. Within two months they were again planning their wedding. Judy knew that it wouldn't be easy if Lou became debilitated and she had to care for him as his condition deteriorated. She told me that she loved Lou more than she could imagine ever loving anyone else. He had a 50% chance of being healthy all their lives and she would take that chance.

But there was another, even tougher decision to be faced: should they have kids? Judy's love for kids had only increased during her two years as a teacher. Her dream of a big family was still strong. However, it had been their discussion of family size that had triggered Lou's decision to leave.

If they had children, it would be before they knew whether or not Lou carried the Huntington's disease gene. If Lou did develop the disease, not only would each child have a 50% chance of developing the disease later in life, but Judy would have to care for her younger children as they grew up, while caring for Lou as he became more helpless. After a month of discussions, including several with a genetic counselor and with their pastor, Lou and Judy decided not to have children. I know that this decision was the most difficult one Judy had ever made.

I decided that I had to learn more about this disease. Something must be known about what happens to the nervous system of a Huntington's victim, and why. Most studies have found what appears to be programmed premature, localized nerve cell death in several areas of the brain. In advanced cases, the weight of the whole brain may be decreased by 20 to 30%. The underlying biochemical defect that results in selective, premature neuronal cell death is not known. The dominant pattern of disease inheritance suggests a single abnormal gene product, possibly one of the neurotransmitters, a group of small molecules which transmit impulses between neurons. But most studies have found normal amounts and functions of these neurotransmitters in the brain. Until the cause of Huntington's disease is found, specific drug therapies cannot be designed (Martin, 1984).

ATTEMPTS TO DEVELOP A TEST FOR HUNTINGTON'S

If no treatment is available, the next best thing would be a test to determine if a person at risk is carrying the Huntington's gene. The uncertainty of whether or not Lou carries this gene and will develop the disease has been the most disruptive force in his life. But without knowing which of the 100,000 genes distributed over the 46 human chromosomes, carries the Huntington's trait, there is no place to start.

Years of research have been spent trying to find a "polymorphic" protein that might be inherited with the disease. Polymorphic means that even though a particular protein is made by every human being, there are slight but detectable differences in the form of that protein. These differences are inherited: if you make "form A" of this protein, it is because one of your parents carried the gene coding for "form A" and passed it on to you.

But how could a protein which is not related to the disease lead to detection of the disease gene? If the gene which codes for this protein is located on the same chromosome, close to the disease gene, it would be inherited with the disease gene. Inheritance of "form A" of this protein would be a genetic "marker" for inheritance of the disease gene. Years of frustrating research were spent without finding a polymorphic protein marker for the Huntington's disease trait.

Proteins are not the only biomolecules which are polymorphic. Deoxyribonucleic acid (DNA) is also polymorphic. In a landmark 1983 paper, Gusella *et al.* reported the location of a polymorphic DNA "fragment" which was inherited with the Huntington's disease gene. The method used to detect this polymorphism is a bit complicated, but to understand the potential and problems of this technique, it is important to understand the principles on which it is based.

Locating Important DNA Sequences

DNA carries the information needed for all the functions of cells and the organisms they compose. DNA is deceptively simple for such an important job. It is composed of two strands of only four building blocks, called bases: adenine (A); thymine (T); guanine (G); and cytosine (C). The "sequence" in which these bases are put together makes one part of a DNA molecule different from any other in that individual.

The DNA of a human being is inherited, 23 chromosomes from the father and 23 analogous chromosomes from the mother. To examine an individual's chromosomal DNA, it is extracted from his white blood cells and cut into discrete pieces with a restriction enzyme. Restriction enzymes cut the DNA every place they find a specific sequence (e.g., at the *in A*AGCTT). The same enzyme will cut two individuals' DNA at most of the same places, resulting in DNA fragments of identical size. However, a single base difference in this cutting sequence will destroy, or add, cutting sites. The result is DNA fragments which are one size in one individual and another size in another individual. Just like the inherited polymorphic protein varieties, these restriction fragment length polymorphisms (RFLPs) are inherited.

An RFLP for Huntington's

RFLPs are detected by cutting an individual's DNA with a restriction enzyme and then picking out a few of the DNA fragments. After cutting, the individual's DNA fragments are separated by size in an electrical current. Specific fragments are located with a DNA "probe," a piece of human DNA previously removed from a chromosome, inserted into a bacterial virus, and maintained in the laboratory. The probe DNA is labeled with radioactive phosphorus (^{32}P). When it is added to the separated fragments, it will locate and bind only to its DNA sequence. The result of such an experiment is shown schematically in Figure 1. (Don't be alarmed if you don't completely understand Figure 1 on the first look; parts of it will be clarified in the following paragraphs. If you want to understand the details of how all this works, just take your time going through the next couple of sections, referring back and forth between text and figures as necessary. Or, if you don't care about the details, you can skim lightly over these sections, just watching for the important conclusions.)

Gusella *et al.* (1983) collected 12 DNA probes which were able to detect

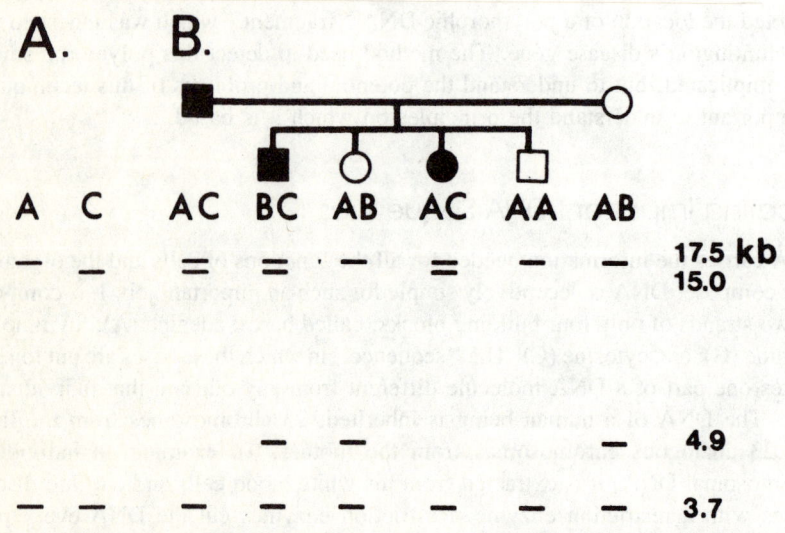

FIGURE 1. RFLP pedigree of a family with Huntington's disease. Males (□) and females (○) are indicated in the pedigree on the top line. Those affected by Huntington's disease symptoms are represented by closed symbols (■ , ●). DNA from each individual's white blood cells was cut with a restriction enzyme and the fragments were separated in a gel by an electrical current. The G8 DNA was labeled with radioactive phosphate (^{32}P) and allowed to find its DNA sequence among the fragments. The excess, unbound ^{32}P-DNA was washed away. The bound radioactivity was detected by exposure to X-ray film which is sensitive to the β-emissions from ^{32}P. The pattern shown in Panel A is hypothetical. If an "AC" individual's paternal chromosome could be separated from his maternal chromosome, one would be "type A" and the other "type C." In reality, this individual would have the composite A + C pattern seen in the first column of Panel B. The key for deciphering these RFLP patterns is presented in Figure 2. This figure was assembled for illustrative purposes from Gusella et al. (1983) and Gusella et al. (1984).

RFLPs between individuals. They set out to examine the DNA from members of a large American family with Huntington's disease. The task was mammoth: 100,000 genes vs. 12 probes. Incredibly, one of these probes, G8, from Tom Maniatis at Harvard (Lawn et al., 1978) detected a RFLP which appeared to be inherited with the Huntington's disease trait!

To test the association between the Huntington's disease trait and the G8 RFLP, a second family from a unique community of Huntington's disease carriers on the shore of Lake Maracaibo in Venezuela were tested. This pedigree included 3,000 inhabitants, many of whom had inherited the disease gene. It appears that a European sailor in the middle 1800s brought the disease, which has been passed from generation to generation. All Huntington families for which data is available can be traced to Europe, probably the result of a single original mutation which has now been passed down through many, diverse family trees.

Problems for a Huntington's Test

Now that a RFLP which is inherited with the Huntington's disease trait has been found, we should be able to test Lou to determine if he truly has anything to worry about, right? Wrong, very wrong. The G8 RFLP was used as a predictor in the two large families that Gusella et al. (1983) tested, but only because they developed an extensive family pedigree for the marker, and the disease. To understand why the determination of Lou's G8 RFLP pattern, alone, cannot predict his future, it is important to understand more about how the RFLP system works and why a large family pedigree, and some luck, is required.

The first complicating factor is that each individual has two copies of each chromosome, one inherited from his mother, the other from his father. *If* these two sets of chromosomes could be separated before the RFLP analysis, the DNA fragment pattern might look like that shown in Figure 1a. The father's "type A" pattern would yield 17.5 and 3.7 Kb (kilobase: 1,000 bases) fragments, while the mother's "type C" pattern would yield 15 and 3.7 Kb fragments. Unfortunately, the father and mother chromosomes cannot be separated before the RFLP analysis. The actual result of such an analysis is a composite of the type A and C patterns, as found in the first column of Figure 1b. These composite patterns can be interpreted logically, and each family member can be assigned a "type" designation, as described in Figure 2.

Knowing which family members were affected by Huntington's disease, and determining that the G8 RFLP "type C" was always found in those individuals with the disease (Figure 1b), led Gusella et al. (1983) to the conclusion that the "type C" pattern was associated with the disease. But this conclusion is only valid for that particular family. The "type C" pattern is also found in 25% of the non-Huntington's population. These normal people have the normal "Huntington's gene" associated with the G8 marker. In other words, the association of the G8 marker with the disease gene is only informative in families where the G8 marker

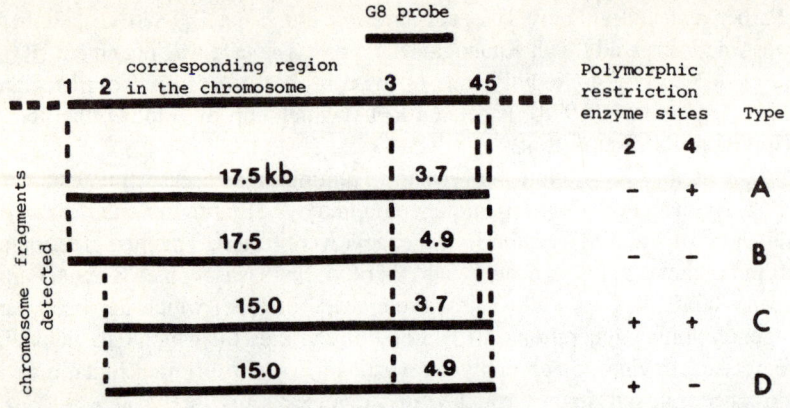

FIGURE 2. RFLPs of the G8 sequence found in human DNA. The G8 probe DNA is presented at the top of the figure, directly over its sequence in human DNA. The second line shows the possible restriction enzyme cleavage sites (numbered 1–5) which have been found in the corresponding region of human DNA. The key is that cleavage sites 2 or 4 may or may not be present in each copy of this DNA from a particular person. The presence or absence of cleavage at site 2 or site 4 will determine the size of the fragment. G8 will bind to any fragment with which it shares nucleotide sequence. In other words, even the large 17.5 Kb (or 15.0 Kb) fragments from this human genome segment will be recognized by G8 because it shares the base sequence at its right and with G8's left end.

The 4 possible combinations of genome fragments are presented in the bottom 4 lines. A summation of the presence or absence of restriction site 2, or site 4 is shown on the right. Each combination is assigned a "type" (A–D). Since each individual has two copies of each chromosome, his RFLP type can be described by 2 letters (AA, AB, BC, etc.), as shown in Figure 1. (Modified from Gusella et al. (1983) and Gusella et al. (1984))

has been rigorously associated with the disease trait. This association can only be accomplished with a large family pedigree, such as the ones in Gusella *et al.* (1983).

Unfortunately, a few exceptions were found to the original rule that the disease is associated with only one G8 RFLP type in one family. A "switch" was found in a member of one of the original families (Gusella *et al.*, 1984), and two other switches were found in two other families (Folstein *et al.*, 1985). These switches result when a part of the chromosome from one parent recombines with the same part of the chromosome from the other parent. The result is a chromosome with one end from the father and the other end from the mother, as shown in Figure 3. The farther the disease gene and the G8 marker are from each other on the chromosome, the more likely an individual with a switch will eventually be found.

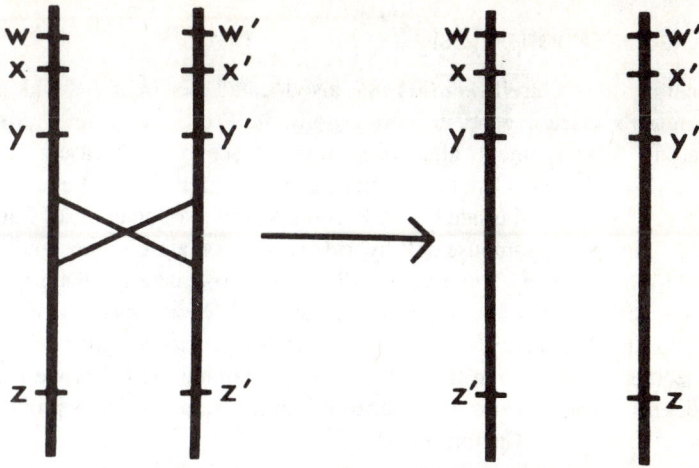

FIGURE 3. Switching (recombining) DNA between chromosomes. Recombination between the maternal and paternal copy of the same chromosome in an individual is a normal occurrence. The site at which two chromosomes recombine is random. Therefore, the likelihood that recombination will separate two markers is related to the distance between them. A long distance between the markers, as between X and Z, results in a high likelihood that recombination could take place between them and separate them. A short distance between the markers, as between X and Y, results in a low likelihood that recombination would occur between them.

If the recombination shown here took place, the DNA fragments recognized by the G8 probe (Z) would be separated from the Huntington's disease gene (X). If a new DNA probe were found which is inherited with the disease gene in his ancestors as well as in his own "recombined" DNA, the new probe would either have to be closer (Y) to the disease gene (X), or on the other side of it (W).

After their first study, Gusella *et al.* (1983) estimated that the disease gene and the G8 marker were within several million bases of each other. But the number of switches which have been found increases that estimate. In fact, this finding was foretold in the original study, since the Venezuelan family disease gene was associated with the "type C" pattern while the American family disease gene was associated with the "type A" pattern.

Nevertheless, the G8 marker is strongly associated with the Huntington's disease gene. But it is not close enough to make a completely accurate diagnosis, even in the large well-studied families (Folstein *et al.*, 1985). G8 will be useful in searching for closer markers and eventually finding the disease-causing Huntington's gene. G8 has already narrowed the search to 0.1% of the 3 billion bases of total human DNA.

Prospects for a Better Test

The original report of Gusella *et al.* (1983) also located the G8 probe, and therefore the Huntington's disease gene, to chromosome 4. Presently, several groups are working along chromosome 4, attempting to get closer to the disease gene. James Gusella (at Massachusetts General Hospital) is looking for a second marker for the disease gene, on the opposite side from G8. John Wasmuth (University of California, Irvine) is making human–mouse cell hybrids which contain chromosome 4 as the only human chromosome. These cells will be used to generate a battery of chromosome 4 specific probes for further linkage studies. David Schwartz and Charles Cantor (Columbia University) have developed techniques for isolation of very large DNA molecules. Such a large DNA molecule, containing the G8 probe sequence, may be identified and sections of it cloned to make a series of DNA probes to test on the family pedigrees (Cantor, 1984).

Ironically, the few family members in the original studies where chromosome switches were found will be useful in narrowing the search to a smaller area, closer to the gene location. If a new marker is found which associates with the disease gene in these cases, it must either be closer to the disease gene than G8 is, or on the other side of the disease gene, as described in Figure 3.

Eventually, these techniques will provide a physical map of a few million bases in the vicinity of the Huntington's disease gene and a set of markers for all the DNA in this region. RFLPs will continue to be used to find markers whose inheritance matches more and more closely the disease gene in a series of families.

These approaches should be able to lead researchers to within 500,000 bases of the disease gene. The exact location of the Huntington's disease gene runs into a roadblock at this point. The DNA polymorphisms which have been so useful in getting this close are now a hindrance. If the entire DNA sequence of this region were determined, there would be several thousand base differences between individuals. Which one represents the Huntington's disease mutation? Without knowledge of the protein produced by this gene, there is no direct way to find the gene.

How to Locate a Gene in a Haystack

The only thing known about the Huntington's disease gene is that it causes degenerative neurologic disease in afflicted adults. How might the disease gene be isolated using this function? It is now possible to insert foreign genes into mice by injecting DNA into fertilized eggs. About 10% of the time the DNA inserts into a mouse chromosome and becomes part of that chromosome. As the embryo divides, so does the foreign DNA in the chromosome. The gene is even passed on from generation to generation. Different sections of chromosome 4 from a Huntington's disease patient could be injected into fertilized mouse embryos, and mice which have incorporated the human DNA could be selected and observed. Since the Huntington's disease gene is dominant in a human, it might also be dominant and cause a degenerative neurological disease in a mouse. An advantage to using the mouse is that it reaches mid-life, the time of Huntington's disease in humans, within a year after birth. Of course, there is no way of knowing whether this human gene would function in a mouse. There is some hope that it might, since brain proteins of many animal species share structural similarities.

If a piece of DNA from a Huntington patient would cause disease in a mouse, this DNA could be divided into smaller and smaller sections and injected into mouse embryos until the smallest disease-causing piece is isolated. The analogous piece of DNA from normal and Huntington's disease patients could then be completely sequenced and compared for a common mutation in the patients. These studies will take many years to perform, but there is hope that they might succeed.

Precise location of the disease gene would be the ideal starting point for the ideal genetic test. That goal may still be far off. However, an RFLP marker closer than G8 to the disease gene should be found in the next few years. It will then be possible to make a test for the disease gene available to many of the potentially affected individuals, providing the individuals have DNA from enough affected and unaffected relatives to develop a pedigree. As an important step toward preserving this information, a national DNA bank has been established at the Indiana University Medical Center, Indianapolis, to store specimens from family members whose health is jeopardized. Blood samples are drawn and the DNA is extracted and preserved. These DNA samples will be useful in the future to establish family pedigrees.

TO TEST OR NOT TO TEST

Of course, if a test is developed, many people with Huntington's disease in their family will choose to be tested. Recent studies of people in this situation have revealed that 75% would like to be tested, to know if they will fall victim to Huntington's disease (Koller and Davenport, 1984). Such a test would have to be extraordinarily accurate. It would have to be carried out in duplicate, in two laboratories, and only reported to the patients when both tests agree.

Lou has been forced to live with the uncertainty. Two years after he married Judy, he began noticing the early signs of Huntington's disease: stumbling for no reason, dropping a glass here and a dish there. He became convinced that the disease was starting. He didn't tell Judy, for fear of upsetting her. Finally he felt certain. Lou had watched his mother deteriorate and he had vowed not to let his life end in the same slow, helpless way. One Saturday afternoon, he set out for the grocery store and drove off the road, into a telephone pole. Lou survived with a broken arm and leg. He told Judy that he knew that the symptoms had begun. A neurologist checked him out but found no signs of the disease. But Lou had entered a deep depression. He had to leave his job. It took Lou a year to regain his equilibrium.

Lou would take a Huntington's disease gene test, if it were available. A negative result would alleviate his fears and might well have saved him from his accident and slow recovery. However, a positive result confirming that he has the

Huntington's disease gene would be tantamount to a life sentence. How would that affect the quality of his life? On balance, would it be worse than not knowing? Most people surveyed felt that if they were to be affected, they would like to know in order that they might make plans for their own care and the care of their family.

A test for Huntington's disease would be a unique situation. It would be a test for a disease which has no treatment, no specific determination of when it will begin, and no hope of survival. Even a diagnosis of cancer offers hope in treatment regimens. The only hope that a Huntington's patient might have would be for a new treatment. If such a test result is reported to a patient, it should be by a health care team of an informed neurologist, geneticist, psychiatrist, and a social worker (Wexler *et al.*, 1985). A positive test result might lead to depression or suicide.

A test for the Huntington's disease gene might also be required by employers or insurance companies to avoid taking bad risks (Kolata, 1986). The cost of caring for a Huntington's disease patient can be very high. However, it hardly seems fair to deny my brother-in-law employment, or medical care when he may really need it.

Lou is back to work in a new but smaller architectural firm where he is doing more than ever, and loving it. Judy is still teaching and loving her first-graders. They have struggled with their decision not to have children for five years. They are still not sure whether or not Lou has the disease gene. Of course, if Lou and everyone in his situation would remain childless, Huntington's disease would be gone in a single generation. The chances that Lou and Judy would have an affected child are 25%: the average of a 0% chance (if he has not inherited the disease) and a 50% chance (if he has inherited the disease gene from his mother). With their lives back together, and a resolution to live life however it comes, Lou and Judy did decide to have a baby. Last spring my sister Judy gave birth to my favorite niece. They are taking a chance. But they are so happy with Suzie that right now it all seems worthwhile. Their new decision is to have only one child.

If a test had been available, perhaps Judy could have undergone amniocentesis early in pregnancy. DNA from the white blood cells of her fetus could have been tested for the Huntington's disease gene. If it were found, Judy would have had the option of abortion. Of course, if the fetus did have the disease gene, it would mean that Lou also had the gene. He would learn in a roundabout, but just as accurate, way what fate he had to look forward to.

None of these questions and problems have easy answers, but this work in molecular biology is building a road which could open new options for Huntington's disease families. Eventually, this work will lead to the discovery of the Huntington's disease gene, its normal function in the nervous system, and its abnormal function in disease. Similar research with Duchenne's muscular dystrophy, cystic fibrosis, and other genetic diseases has located DNA markers which are inherited with these diseases. In fact, the actual gene which appears to cause Duchenne's muscular dystrophy has recently been located (Monaco *et al.*, 1986). These workers found that several Duchenne's patients had lost a specific piece of chromosomal DNA. Molecular biology is also being applied to other hereditary disorders like neurofibromatosis, familial Alzheimer's disease, and manic depressive illness.

Meanwhile, I am a proud uncle of a most beautiful niece. Lou and Judy and one-year-old Suzie are living a normal life in a busy household. Will Lou and Suzie take the test if it becomes available? They haven't decided. Perhaps Suzie should make her own decision when she is old enough. Perhaps someday, when we learn what this gene is and does, and how it changes in the disease, a treatment can be developed to alleviate the symptoms. Maybe even gene therapy will someday be available to replace the disease gene. It is unlikely that these therapies will be developed in Lou's lifetime. I hope he will never need them.

POSTSCRIPT

Since this article was written, three DNA probes have been identified which are closer to the disease gene than G8 (Gilliam et al., 1987. *Cell* 50:565–571; Gilliam et al., 1987. *Science* 238:950–952; and Wasmuth et al., 1988. *Nature* 332:734–736, 1988). Two of these DNA probes are so close to the disease gene that no test subject has been found whose DNA has recombined between the marker and the disease gene (see Figure 3). Moreover, in the area detected by the probes, there are many more "polymorphisms" (differences in restriction-enzyme cutting sites among individuals). These markers will be informative in more families, they will allow more accurate diagnosis in these families, and they will allow diagnosis in smaller families. In addition, they are probably within 1,500 kilobase pairs of the Huntington's disease gene and provide much better starting points to identify it.

REFERENCES

Cantor, C.R. 1984. "Charting the path to the Gene." *Nature* 308:404–405.

Folstein, S.E., Phillips, J.A., III, Meyers, D.A., Chase, G.A., Abbott, M.H., Franz, M.L., Waber, P.G., Kazazian, H.H., Jr., Conneally, P.M., Hobbs, W., Tanzi, R., Faryniarz, A., Gibbons, K., Gusella, J., 1985. "Huntington's Disease: Two Families With Differing Clinical Features Show Linkage to the G8 Probe." *Science* 229: 776–779.

Gusella, J.F., Tanzi, R.E., Anderson, M.A., Hobbs, W., Gibbons, K., Raschtchian, R., Gilliam, T.C., Wallace, M.R., Wexler, N.S., Conneally, P.M., 1984. DNA Markers for Nervous System Diseases. *Science* 225: 1320–1326.

Gusella, J.F., Wexler, N.S., Conneally, P.M., Naylor, S.L., Anderson, M.A., Tanzi, R.E., Watkins, P.C., Ottina, K., Wallace, M.R., Sakaguchi, A.Y., Young, A.B., Shoulson, I., Bonilla, E., Martin, J.B., 1983. A Polymorphic DNA Marker Genetically Linked to Huntington's Disease. *Nature* 306:234–238.

Kolata, G. 1986. Genetic Screening Raises Questions for Employers and Insurers. *Science* 232:317–319.

Koller, W.C. and Davenport, J., 1984. Genetic Testing in Huntington's Disease. *Annals of Neurology* 16:511–513.

Lawn, R.M., Fritsch, E.F., Parker, R.C., Blake, G., Maniatis, T., 1978. The Isolation and Characterization of Linked δ (delta) and B-globin Genes From a Cloned Library of Human DNA. *Cell* 15:1157–1170.

Martin, J.B. 1984. Huntington's Disease: New Approaches to an Old Problem. *Neurology* 34:1057–1072.

Monaco, A.P., Neve, R.L., Colletti-Feener, C., Bertelson, C.J., Kurnit, D.M., Kunkel, L.M., 1986. Isolation of Candidate cDNAs for Portions of the Duchenne Muscular Dystrophy Gene. *Nature* 323:646–650.

Wexler, N.S., Conneally, P.M., Houseman, D., Gusella, J.F., 1985. A DNA Polymorphism for Huntington's Disease Marks the Future. *Arch Neurol* 42:20–24.

Patrick Collins

SPACE TOURISM —

THE DOOR

INTO THE

SPACE AGE

JUNE 1988

DREAMERS about human expansion into space have tended to think in rather lofty terms, such as exploring new frontiers and insuring the survival of our species against planetary disasters. For it to happen in reality, though, someone must be persuaded to finance the effort; and the motivations for doing that may turn out to be a bit closer to home. Imagine the Moon, for example, as a tourist attraction. . . .

Patrick Collins worked as a consultant to the European Space Agency from 1979 through 1981, while earning his doctorate on the economics of satellite solar power stations. Since 1983, he has taught managerial economics in London University's Imperial College of Science, Technology, and Medicine and has continued his research on the commercial prospects for the space industry. ■

THE idea of tourism in space plays a part in several well-known science fiction stories: Robert Heinlein's 1957 story "The Menace from Earth" is about a rich tourist learning to fly in a large canyon on the Moon which has been covered over and filled with air for recreational flying—a realistic possibility for the 21st century. Arthur Clarke's 1961 novel *A Fall of Moondust* concerns the misadventure of a party of lunar tourists, while Joanna Russ's 1968 novel *Picnic on Paradise*, set much further into the future when interstellar travel is common, concerns a party of tourists who get caught up in a civil war while visiting a distant planet. More recently, "Galactic Tours," by David Hardy and Bob Shaw, illustrates a whole range of tourist possibilities, from skiing on Europa (one of the moons of Jupiter) in our solar system, to weird and exotic possibilities in other star systems. And one day humans *will* go skiing on Europa—at least a few early explorers will within the 21st century. Most recently, Ben Bova devoted several pages of his "Moonbase Orientation Manual" (Part II, July 1987 *Analog*) to tourism on the moon 50 years from now. With continuing world economic growth, the longer-term prospects for space tourism are clearly limitless.

However, the *beginnings* of space tourism will be very much nearer home, quite literally, for the first destination will be a mere 200 miles away—in low Earth orbit. And although less exotic than the longer-term prospects, visits to nearby space will be hardly less attractive. Despite the seriousness of their training, and their tight work schedules when in orbit, everyone who has experienced it agrees that space travel is *fun*. Sally Ride, the first English-speaking woman astronaut, summed it up when she said of her first flight in the Space Shuttle: "It was great fun . . . and I guess it will be the greatest fun I ever have." U.S. Senator Jake Garn even wangled himself a flight. Irrespective of any scientific or political value that their flights may have had, none of the astronauts would have missed them for *anything*.

More than being fun, a visit to low Earth orbit is also *fascinating*. The absence of gravity introduces novelties into every activity, making the experience endlessly interesting. Washing, dressing, eating meals, and other ordinary activities are all transformed in zero gravity. Even just moving around is so different that the Skylab astronauts reported that they could never resist making acrobatic movements, somersaults, spins and so on, when they had to move some distance. In addition to this, the view of Earth from low orbit is literally breathtaking, both by night when the globe flickers with lightning storms and polar aurorae, and by day when the ever-changing terrain below is dazzlingly clear. Photographs and films of the view are beautiful, but to experience it for real is apparently stunning. Astronauts in Skylab spent *hours* on end watching the Earth through the porthole whenever possible. The absence of air also, of course, provides perfect conditions for observing the Moon, Sun, planets, stars and nebulae.

Even more than this, however, and perhaps most important for anyone (including many science fiction readers) with some feeling for the *immense* future stretching ahead of the human race, as we explore in turn the solar system, the nearby stars, the whole milky way galaxy (and eventually of course, even other galaxies), these first tiny steps above the Earth's surface have extraordinary evolutionary significance. To visit low Earth orbit at this point in history is to feel yourself as part of our species just peeping out into space, as the first protoamphibians must have peeped out of the prehistoric seas at the land covered in primitive plants. As such, the experience of seeing the sky darken into the blackness of space as you climb above the Earth's atmosphere carries such emotions of excitement, awe, and inspiration that for many people a trip to orbit will be an absolute *must*, an almost magical event, a once-in-a-lifetime, modern-day pilgrimage.

Even for those not lucky enough to have this perspective, a visit to low Earth orbit will nevertheless be uniquely entertaining—and the public already understands this. Astronauts have been children's heroes for decades (despite their limited activities since the Apollo program), while a recent survey carried out for the American Express Company in the UK found that more than 50% of those under 45 years old and more than 60% of those under 25 would like a holiday in space if it was available. Even the most experienced traveler who has "been everywhere," from the Caspian Sea to Jamaica, from the Taj Mahal to Alaska, has never done anything remotely like taking a holiday in space.

However, although these early space tourist services will clearly become available long before more exotic services such as trips to the Moon, many of you may feel that even trips to low Earth orbit are not going to be available on a commercial basis within your lifetimes. This, I claim, is *wrong*, and much of the rest of this article will argue the case that it is feasible for space tourism to begin *this century* (yes, during the 1990s), and that once it starts it will grow rapidly and the price will fall so that within 20 years, although still expensive, orbital trips will become widely available for people prepared to save their holiday budget for a few years.

The key, of course, is to bring the costs down within customers' reach, and unfortunately the U.S. Space Shuttle is a step in the wrong direction: Due to the conflicting political requirements placed on its design, it costs *more* to launch a ton of payload than the expendable Saturn V of the 1960s! However, there are two design approaches for reusable launch vehicles using existing technology which could bring down the cost of a short trip into space low enough to establish a profitable industry (see below). Nevertheless, in order to reduce costs sufficiently, the turnover of passengers must be high enough to gain the maximum economies of scale—which will depend both on the price and on the popularity of what is being offered. And I believe that when it becomes commercially available, the possibility of paying a short visit to a "hotel" in low Earth orbit will be extremely popular, even at a price of thousands of dollars, because of the unique range and variety of entertainments that will be available.

ZERO GRAVITY ENTERTAINMENTS

In addition to the extraordinary views to be had from an assortment of portholes, panoramic windows, viewing domes, and observatories, holidays in orbiting "hotels" will provide *zero gravity*. Among other activities, this provides the opportunity for human-powered flight in large gymnasiums, using fabric wings attached to the arms and tails attached to the ankles. Just *learning* to fly will be fascinating in itself, but flying will also provide further scope for many new leisure and sports activities—racing, aerobatics, chase games, dancing. For those interested in the details, flying in zero gravity will not be the same as flying on Earth since there will be no need to generate lift: objects continue moving in a straight line unless they experience an external force. However, wings and tail will be needed for propulsion, steering, and stopping—so having accelerated to the speed you want by flapping your wings, turning, swooping, and coming to a stop will be much like that of a bird. For instance, altering the angle of attack of one or both wings will allow you to roll, climb and dive at will, while raising your tail segments (attached to your ankles) will tip your body "down" perpendicular to your flight direction, and flapping your wings from back to front (while keeping your balance!) will bring you to a halt. There's clearly plenty to learn!

Zero gravity water sports in a large, gymnasium-sized room will also offer many new experiences. The behavior of water in zero gravity is dominated by surface tension, and a swimming-pool sized "piece" of water could be broken up into dozens of different sized "pieces" offering different attractions: It would be possible to dive right through large pieces and emerge from the other side, while swimming would have the novelty that bodies at the surface would float very high, whereas "underwater" you would not rise to the surface spontaneously. (For safety, people may wear small bottles of compressed air with a mouthpiece to allow them to breathe when underwater.) Armfuls of water could also be thrown as a means of propelling yourself around the room, while smaller pieces the size of tennis balls would be as handy as snowballs! A room containing many "pieces" of water would provide an interesting environment for hiding and chasing games. A large, slowly rotating, cylindrical swimming chamber would also enable people to swim around the inside curve in low pseudo-gravity for exercise, as well as to dive out and float in the central air space.

Zero gravity also provides extraordinary scope for completely new ball games, chase games, and games of skill. Acrobatics would become a different, more leisurely, high precision art form. The use of simple air thruster packs would allow people to maneuver in three dimensions without effort, or to "dog-fight" with each other. Just learning to make gentle, controlled movements while preserving your balance will be interesting, and sleeping with a partner will have its novelty, for instance in "rendezvous and docking," as well as its advantage—no more arms going to sleep! The list of enjoyable uses of zero g is as long as your imagination.

For those wanting less energetic pursuits (or those temporarily exhausted) zero gravity also offers fascinating possibilities for demonstrating physical phenomena not possible on Earth. Water and other liquids behave quite differently; they can be formed into "ropes" or rotating donuts, or expanded with bubbles that don't rise to the surface and burst. This would make it possible to bake "ultralight" cakes or to make objects out of metal "foam," while the unconstrained behavior of magnetic materials and spinning objects can also be shown in zero *g*. Again, the possibilities are endless.

On a different note, orbiting botanical gardens would hold enormous interest as they revealed the exotic ways in which different plants adapted to a zero gravity environment. Many will grow much larger than they do in the 1 *g* gravity field on Earth, and they will grow in much weirder ways without the gravity vector to guide them. A 3-D ramble through a zero *g* hanging jungle, complete with "waterfalls" and pools, will be a possibility at a later stage! It is easy to get carried away musing on the delightful possibilities that are going to be available one day, but we must come back to Earth if we are to seriously consider how soon these ideas may be feasible.

PROSPECTS FOR NEAR-TERM SPACE TOURISM

There are several reasons for being optimistic about the feasibility of a commercial space tourism industry starting this century. The first reason is that *technically* the project lies well within what is possible today. In most respects the technical requirements of both the necessary launch vehicles and the orbital accommodation units lie *within* the limits of technology that either exists or is currently being developed. For instance, accommodation modules required for an orbiting hotel are much less technically demanding than Spacelab modules: There is no requirement for state-of-the-art laboratory hardware, computing equipment, or telecommunications facilities. Nor are large amounts of power required; nor are accurately controlled attitude or gravitational fields required in a hotel. Much of the astronomical cost of the planned U.S. space station is to be incurred in developing new technology in all these areas.

Tourism merely requires comfortable accommodations and leisure facilities, which require only standard structural modules with lightweight interior partitioning and suitable furnishings. Environmental control and life support, power, thermal control, attitude control, communications, and other systems would need little adaptation from systems that already exist. The only significant new developments are the need for plenty of windows in every module and semi-autonomous environmental control systems with multiple redundancy throughout the facility—and even for these the technologies are already fully developed. The single respect in which space tourism operations would need to be innovative is in achieving standards of safety of both launch vehicles and accommodation facilities similar to civil aviation standards. This is impossible to achieve with current launch vehicles, which are expendable or partly expendable, but the use of fully reusable launch vehicles will enable reliability to reach the level of civil airline operations through step-by-step development.

A second reason for optimism about the prospects for space tourism is the fact that there is enormous scope within the space industry for reducing costs. Space projects today typically involve politically determined developments financed exclusively by governments, followed by production of small numbers of the end-product. This is quite unlike normal commercial industries, and among other consequences costs remain extremely high. As an example, while U.S. industry was developing the Space Shuttle during the late 1970s and 1980s, Europe and Japan broadly repeated U.S. work of the 1960s with their development of the HM7 and LE5 cryogenic rocket engines (more or less duplicating the Pratt & Whitney RL10 engine), and they are currently developing the HM60 and LE7 engines (which roughly duplicate Rocketdyne's J2 engine). The resulting low utilization of all these engines ensures that their costs remain extremely high, despite the fact that they are inherently simpler than jet engines.

There have also been some spectacular examples of cost reductions: A company in the U.S. auto industry was recently invited to manufacture some nickel-hydrogen batteries used in NASA satellites. By applying their commercial expertise in cost reduction, they reduced the production cost from $25,000 to less than $1,000, and they expect it to fall eventually below $400! Likewise the cost of food on the Space Shuttle is a fraction of the Apollo astronauts' food bill. Much of it is now bought from supermarkets, instead of being developed from scratch.

In recent years the intensity of competition between manufacturing industries worldwide has rocketed—particularly in the speed of new product and model development, increased product quality and reliability, and speed of reaching mass production. These features have become particular hallmarks of Japanese manufacturing industry, and American and European companies are having to rapidly learn new forms of organization and engineering professionalism in order to catch up. The diffusion of these qualities throughout manufacturing industry is, of course, exactly what is required to reduce costs in the space industry—in which the Japanese are also making rapid advances.

A third reason for optimism is the very large scale of the potential demand for space tourism services. As already mentioned, opinion polls have found that more than 50% of the population would take a trip into space if given the opportunity, but the significance of this figure is greatly increased by the fact that it is based only on curiosity on the part of the public. When better informed about the range of interesting possibilities described above, an even higher proportion of the population would be likely to favor a short "holiday" in space. Perhaps most important, however, once space tourism starts, a "word of mouth" effect is likely to accelerate its spread. Hearing how much fun it is from a personal friend who's *been* into orbit will probably be one of the strongest inducements to try it!

Although the high level of spontaneous interest in visiting space is very encouraging, we need to know how many people per year would pay a given price. A "demand curve" is the name of the graph used to illustrate this information (Figure 1). So, for instance, if 10 million people per year were prepared to pay $10,000 for a trip to space, there would be no problem: $100 billion of revenues per year would pay for enormous development expenses. On the other hand, if only 1,000 people per year were prepared to pay $10,000, it would clearly not be feasible:

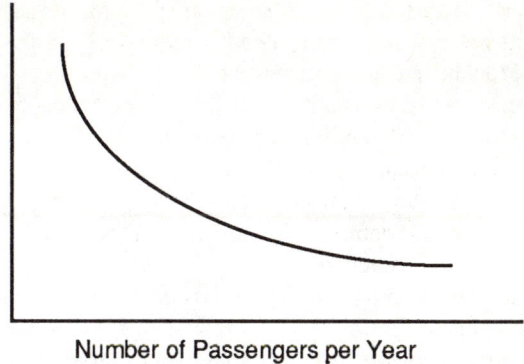

FIGURE 1. Space Travel Demand Curve.

$10 million of annual revenues would be nowhere near enough. So what *do* we know about the demand curve for space tourism?

Market surveys in the U.S.A. suggest that, initially, a few thousand people per year would be prepared to pay between $50,000 and $100,000 for a short trip into low earth orbit, as proposed by the U.S. travel company Space Expeditions. (Their "Project Space Voyage" comprises a seven-day residential stay in a hotel-cum-training-facility on Earth, followed by a twelve-hour orbital flight.) Such a level of demand should be enough to justify initial production of the "Phoenix" passenger launch vehicle designed by Pacific American Launch Systems Inc. Given the likelihood of sales of the "Phoenix" to other customers, some hundreds of millions of dollars of annual tourism revenues should, after deducting operating costs, be able to pay off approximately $1,000 million of initial investment. The project would be a *cinch* if given even a fraction of the billions of dollars of government subsidies provided to the Space Shuttle or Europe's Ariane launcher.

It is not known how many people might pay between $20,000 and $100,000, and obtaining reliable estimates will require well-prepared market research. However, reasonably convincing arguments can be made that as many as one million passengers per year would pay $10,000 for a short stay at an orbital "hotel." This level of traffic (several hundred times current launch rates) would provide major economies of scale: One million passengers, each spending, say, three days at an orbital hotel, would need launch vehicles to transport 20,000 passengers to and from orbit each week, as well as simultaneous accommodation for 10,000 people in orbit, or perhaps 40 hotels for 250 guests each. Such hotels might initially comprise some 30 Spacelab-like modules (50' long by 14' in diameter), plus perhaps six much larger modules made from converted Space Shuttle external tanks (150' by 28' in diameter). The required production runs of 1,200 smaller modules and 240 larger ones, plus perhaps 100 fully reusable launch vehicles, are far larger than have been achieved with space hardware hitherto, and would make a major contribution towards achieving the target of $10,000 per guest.

A fourth factor suggesting that commercial space tourism can be expected to develop rapidly (given that these other conditions make it profitable), is the large scale, the rapid growth, and the competitive vigor of the foreign travel industry today. With approximately one million tourists flying abroad *every day*, and a turnover of more than $100,000 million per year, the industry has an annual growth rate of around 5% in real terms (so it will double again by the end of the century). In seeking to generate the large increase in launch traffic that is potentially available from space tourism, intense competitive pressure will be exerted on launch costs, bringing about major cost reductions.

Although the popular image of tourism is of a rather "lightweight" business, comprising mainly simple services such as restaurants and beaches, it has a long track record as a *major* driver of new technological developments—particularly in transportation, telecommunications, and computers. The continuing technological improvements in commercial passenger shipping over the past two centuries, and in commercial aviation this century, have been driven primarily by the requirements of the public as customers, while the need for global booking, ticketing, and foreign exchange systems required by airlines have put continual pressure on computer and telecommunications systems manufacturers to innovate. Thus it would be no more than a continuation of its past history for tourism to play the leading role in knocking another industry into shape, and bringing the technology of space transportation to commercial maturity.

These facts are very encouraging, and provide good grounds for optimism. The space industry is only just beginning to consider space tourism seriously, but all that is needed is a few of the world's *real* entrepreneurs to take an interest—that is, people who can take serious risks with serious amounts of money, on the order of hundreds of millions! If the initial, higher-priced phase begins before the end of the century, with the vigorous growth in demand that could be expected for such a high-profile service, the turnover could reach one million passengers per year within twenty years.

At $10,000 per head, a holiday in space is clearly still not going to be an everyday affair. After all, who has $10,000 to spare? In other words, until the price falls even further to just a few thousand dollars, it will remain a once-in-a-lifetime experience for most people. However, with continuing economic growth, in 20 years' time most middle-income earners *will* be able to afford one trip in their lifetime. With tax efficient savings plans, if you earned 10% compound interest for 20 years, you would need to put away only $1,500 today, or less than $175 per year. Even at 7% it would be only $250 per year, or $160 per year over 25 years. If a space holiday was *sure* to be available in 20 years' time, many people would be prepared to make this sacrifice, while older people would set up plans on their children's and grandchildren's behalf. Lotteries with tickets as prizes are also likely to become popular, and to be widely used by business as promotional schemes. There are thus many reasons why even a relatively high price (in absolute terms) should not prevent the service reaching a wide public—you and me!

PASSENGER LAUNCH VEHICLES

The costs of all operations in space depend critically on the cost of space transportation, and space tourism is no exception. The design of passenger launch vehicles for low cost, "airline" operations is therefore vital, and work is progressing in two main directions. A major constraint in designing a launch vehicle is the need to carry sufficient propellant. The cost of propellant tanks increases with their surface area, and so, since a sphere has the lowest surface area per unit of volume, a nearly spherical shape is attractive in having the lowest mass for a given quantity of propellant. Using something approaching such a shape, it is possible to design a single-stage-to-orbit vehicle that takes off and lands vertically. Several designs were made of such vehicles in the 1960s and 1970s, and more recently by Pacific American Launch Systems, all of which are fairly close to spherical, being roughly conical, blunt-nosed, simple cylindrical structures. Expeditions' cost target for PacAm's "Phoenix" is $50,000 per passenger, which should be feasible once operations have shaken down, around the end of the century.

The second main route in designing a launch vehicle is to take off horizontally using wings to generate lift in the early stages of the flight. This has the advantage that the initial thrust required is only about 25% of the vehicle mass (instead of about 140% in the case of vertical takeoff), but it is possible only for vehicles with a gross liftoff weight of no more than about 500 tons. (The Space Shuttle's weight of around 2,000 tons makes horizontal takeoff impossible.) Two-stage winged vehicles, in which only a small upper stage reaches orbit, are much easier to design than single-stage vehicles, and a particularly promising low-cost approach is the "Spacebus" design of David Ashford in the UK. This is somewhat similar to the West German "Sanger" vehicle, except that it would be cheaper to develop since it would require less new technology. In such vehicles the propellant cost per passenger will be around $3,000, and the other components of cost will fall progressively as the number of passengers handled grows, given full reusability and long vehicle life.

It therefore seems likely that there will be a choice for space tourists between the more exotic vertical takeoff vehicles, in which they will experience up to 3 g acceleration, and a more conventional airline-like launch in 2-stage winged vehicles. At a later date, when the technology has matured, the development and operation of single-stage winged launch vehicles may be commercially justifiable.

SPACE HOTELS

As mentioned above, the requirements for orbiting hotels could be satisfied easily using equipment developed for the U.S. Space Shuttle and space station. The earliest facilities will comprise simply a few cylindrical Spacelab-type modules, and/or refurbished Space Shuttle, external tanks, as proposed, for instance, by the US External Tanks Corporation. It isn't necessary to be able to predict future developments in detail to foresee that as traffic builds up and prices fall there will be continuing demand for more elaborate facilities. As a result, progressively more advanced hotels will be constructed, starting with larger assemblies of different modules, and later including much larger modules launched in component form and assembled in orbit. Subsequently, there will be tethered extensions providing fractional gravity several kilometers above or below the main section. Rotating sections will also be used to provide pseudo-gravity in the style of the classical rotating space stations. Later hotels will also be put in polar orbits, giving wider views of the Earth, as well as scope for interesting transfer trips between facilities. Ultimately, orbital hotels will include "buildings" with dimensions in kilometers, which will provide almost limitless scope for novel environments and entertainments.

There is one proviso to all the foregoing—namely that the demand for space holidays will depend critically on their both being, and seeming to be, *safe*. This will require assurance on three different matters: First, the vehicles and facilities will have to be mechanically safe, requiring the performance of flight test programs to civil aviation standards, as well as the availability, once in operation, of safety devices and rescue vehicles. Second, health risks must be no greater than for other tourist activities. Short-term exposure to zero gravity has been shown to have no damaging effects, but there will be a need for sheltered areas within orbital hotels to provide protection from solar flare particles (for which there are, conveniently, a few minutes' warning). Third, there must be no significant risks from collisions in orbit (primarily with debris). Achieving this will almost certainly require international agreements to reduce debris in orbit, and eventually orbital traffic regulations.

GLOBAL IMPLICATIONS

If space tourism develops as I suggest, it will have a number of important implications for the rest of the world. First, if one million tourists per year are paying $10,000 for an orbital trip in 20 years time, commercial investment is likely to be rapid, leading to a market of perhaps 20 million passengers per year within a further 20 years. In addition, the reduction in launch costs that the growth of space tourism will bring about will make other activities in space commercially viable. For example, the construction of satellite solar power stations (orbiting solar energy collectors with areas of many square kilometers, to be used to transmit gigawatts of electric power to Earth as microwave or laser energy) will be a profitable investment at such launch costs. The demand for energy in the next century from an increasingly industrialized world, given that the use of both nuclear and fossil fuels are likely to face serious constraints, will drive electricity supply industries to exploit such opportunities. This will lead to further progress in space industrialization, including in particular the extraction and processing of metals, ceramics, and other materials from the Moon and asteroids.

Second, of the changes taking place in the world, one of the most important for humanity is economic growth in the developing countries, where living standards are generally very low. This requires the continuation of the well-established pattern of global economic development whereby countries industrialize by progressing from simpler to more complex manufacturing industries. This depends, in turn, on the richer countries continually developing new and more advanced products to balance the developing countries' progressive takeover of more basic industries. Unfortunately, in recent decades there has been insufficient new employment in the richer countries, and growing political pressure to protect their older industries against the more competitive manufactured goods of the developing countries. In this situation, the development of a rapidly growing commercial space industry will create a dynamic new focus for industrial growth and investment in the advanced countries. This will reduce the pressure for protectionism, encourage increased imports from developing countries, and so remove one of the major obstacles to their economic growth.

A further benefit will be to provide commercial demand for many of the most advanced technologies which have been developed primarily for military purposes. By providing an alternative outlet for many of these technological skills, the expansion of a commercial space tourism industry will reduce the need for governments to encourage exports of military equipment which currently aggravate regional conflicts. Thus the development of a proper, commercial, profit-making space industry in the rich countries, initially driven by the currently unsatisfied demand for space tourism, will help to get them off the backs of the developing countries in more ways than one.

A different, but perhaps equally important, benefit of the development of space tourism will be to motivate young people to study technical subjects at school, by creating enthusiasm for engineering and science. This was very noticeable in the U.S.A. during the Apollo Program of the 1960s, and is now urgently needed to redress the drift away from sciences seen in the U.S.A. and Western Europe in recent years.

EPILOGUE

Many science fiction stories, such as Poul Anderson's "Tales of the Flying Mountains," are set in a future in which industrial society has spread through the solar system—and this is surely inevitable provided that we do not destroy ourselves or the environment of our planet. However, no stories provide a convincing description of how this is going to come about. The technological potential for the development of a commercial space industry has existed for nearly twenty years—essentially since the maturing of liquid hydrogen rocket technology in the U.S.A. during the 1960s. However, although the industry has seen further technological advances in electronics and materials since then, it has barely advanced *commercially*, and its revenues still come almost exclusively from taxpayers. And since the space industry isn't *profitable* today, it is a burden on the taxpayer, and suffers all the ills of being a political football.

The space establishment is currently proposing to spend some $80,000 million of taxpayers' money to send some scientists back to the moon, and maybe twice this to send a manned vehicle to Mars. This is like a government of the 1920s proposing to spend several times the turnover of the aircraft industry of the day on, say, building a "national aeroplane" to carry 500 people around the world: it would have been an inappropriate project at that stage of the industry's development, and would have diverted resources away from commercial purposes. What *actually* happened was *much more valuable*: the government subsidized the establishment of a competitive commercial aviation industry (both manufacturers and airlines) by offering guaranteed mail contracts. The objective of flying round the world was, of course, achieved spontaneously in due course, without further taxpayers' funds.

A similar step today, subsidizing the development of true passenger launch vehicles, would cost only a *small fraction* of $80,000 million, would be *much* more popular with the public, and would be *far* more economically beneficial, not least by establishing a commercially self-sustaining space industry. And once the space industry is independently profitable, its destiny will be in its own hands, with no foreseeable limits to growth. As taxpayers, *we* pay for government-funded space programs. It is therefore up to *us* to tell politicians how to spend these funds. If we want space tourism (which appears to be the case), we have only to start pressing for it. Governments could do a lot worse than take a page out of the history books, and guarantee to purchase a minimum number of passenger seats into orbit each year. The idea that the door into the Space Age will be opened by the "human" route of popular curiosity—people taking holidays in space to see what it's like, rather than elitist activities of central government—is an attractive one.

REFERENCES

Ashford, D.M. and Collins, P.Q. (1987), "Orbital and Sub-orbital Passenger Transport: the Key to the Commercialisation of Space," *Proc. 38th IAF Congress*, IAF-87-632.

Bono, P. (1967), "The Reusable Booster Paradox—Aircraft Technology or Operations?," *Spaceflight*, Vol. 9, 379–87.

Collins, P.Q. and Ashford, D.M. (1986), "Potential Economic Implications of the Development of Space Tourism," *Proc 37th IAF Congress*, IAA-86-446.

Collins, P.Q. and Williams, T.W. (1986), "Towards Traffic Systems for Near-Earth Space," *Proc. 29th Colloquium on the Law of Outer Space*, IISL, pp. 161–170.

Hudson, G.C. (1985), "Phoenix: A Commercial, Reusable, Single-stage Launch Vehicle," *EASCON 85*, 151–163, IEEE.

Koelle, D. (1971), "Beta, a Single-stage Reusable Ballistic Space Shuttle Concept," *Proc. 21st IAF Congress*, North-Holland, 393–408.

Koelle, D. and Kuczera, H. (1987), "Sanger. An Advanced Launcher System for Europe," *Proc. 38th IAF Congress*, IAF-87-207.

Rogers, T.F. (1986), "Space Phoenix Project," *UCAR Foundation*, Boulder, CO.

Joel A. Davis

EXPLORING THE

ASTEROIDS

SEPTEMBER 1982

ANOTHER force likely to attract human beings into space is the prospect of greatly enhancing the material resources available for human use. It has long been known that other bodies in the Solar System contain large deposits of things we could use, and asteroid miners have appeared in science fiction for several decades. It will soon be economically practical to pursue such ventures in reality. What, exactly, might we get from the asteroids—and how might we determine in advance which asteroids will be the best sources?

Joel Davis is a freelance science writer whose articles and news reports have appeared in nearly every major popular science magazine in America, including *Astronomy, Final Frontier, Omni,* and *Science Digest.* He is also the author of several books in fields as diverse as biochemistry and the brain, space exploration, the physics of antimatter, and immunology. He is currently working on *The Genome Project* for John Wiley & Sons. ∎

ONE of the first serious suggestions for exploiting asteroidal resources appeared in an article written by the late Dandridge M. Cole back in 1960. More than 20 years later the prospect of mining, capturing, or colonizing an asteroid still seems audacious. It may seem so—but it isn't. Several recent studies suggest that asteroid mining or capturing may be feasible projects before the turn of the century, with a monetary investment comparable to the Apollo program and using current or near-current technology.

However, the economical exploitation of asteroidal resources is based on a big assumption: that the resources we want *are actually there*.

Everything we currently know about asteroids comes from Earth-based observations or studies of meteorites thought to come from certain asteroids. In many ways our knowledge of asteroids is comparable to our pre-1965 understanding of Mars, before Mariner 4 made its photographic flyby of the Red Planet. While ground-based instrumentation is several orders of magnitude better now than it was in the early 1960s, it's a fact that—as an example—almost none of the 2,300-plus named asteroids can even be resolved to a disk with our highest-power telescopes.

(That will change in 1985 with the orbiting of the Space Telescope; yet, still, we'll only be imaging the surfaces of asteroids and gathering data that's limited in scope.)

To actually mount an asteroid-mining operation on presently available information would be a bit like asking the Kennecott Corporation to dig a full-fledged open-pit copper mine in a location that's never been explored with a geologist's hammer.

Just as we photographed and probed the lunar surface with unmanned robots before landing two humans there, so we must examine a number of asteroids up close: first with robot spacecraft and later, if necessary, with human explorers. As we accumulate large amounts of high-quality data on the surface and interior composition of asteroids, we will move closer to making realistic plans for exploiting the resources of these fascinating and (perhaps) valuable chunks of rock.

CURRENT KNOWLEDGE

Asteroids, to start with, are not a homogeneous group of objects. They can be classified by rough (and somewhat speculative) composition and by spatial location.

Dynamical studies of the entry paths of certain meteorites have led researchers to believe they are associated with some asteroids. Going with this assumption (and it is not an unreasonable one), it's been possible to set up a rough compositional classification of asteroids by comparing their reflectance spectra with that of meteorites. Thus we can talk of C (carbonaceous), S (silicaceous or stony), M (metallic), and U (the ubiquitous "unknown") bodies. The C-type asteroids may have substantial amounts of volatile compounds such as water, carbon, and organic molecules. S and M asteroids are not kinky; they may contain sizeable amounts of iron, iron-nickel, and perhaps other metallic elements and compounds.

It's easier to classify asteroids by where they are in the solar system. Doing it that way, they fall into three main classes: the Main Belt asteroids, orbiting between 2.17 and 3.3 AU from the sun between Mars and Jupiter; the Trojan asteroids, occupying the L-4 and L-5 Sun-Jupiter Lagrangian points; and the Earth Approach Objects (EAOs), which occupy orbits that approach or cross that of Earth.

No one's yet found, by the way, any asteroids in the Sun-Earth Lagrangian points, though searches have been made. There are also a few oddballs around; best known is (2060) Chiron, which orbits between Uranus and Saturn. (When a newly discovered asteroid has been observed often enough for its orbit to be plotted, the International Astronomical Union gives it an official number, and its discoverer gives it an official name. Chiron is the 2,060th such asteroid, and discoverer Charles Kowal named it after the centaur in Greek mythology who was the teacher of Hercules.) It could be the first known member of a whole new class of asteroids (the Minor Belt? the Far Belt?).

We know of more than 2,400 asteroids; perhaps 100,000 or more are bright enough to be discovered photographically and they range in size from 1,025 km. ((1) Ceres) down to objects most properly called flying mountains.

Most Main Belt asteroids are pretty well confined to the plane of the ecliptic. The average orbital inclination is 9.7 degrees, and the average eccentricity is 0.15. The immense gravity field of Jupiter does tend to smear things up a bit at the outer edges of the belt. A compositional difference may also exist, depending on location. Asteroids in the inner belt tend to be M or S bodies; those in the outer belt are more likely to be C-type asteroids.

The Trojans, in orbits that form "clouds" around the Sun-Jupiter Lagrangian points, show some evidence of being covered with organic, kerogen-like compounds. They are a long way off, though, and asteroids just as interesting and potentially valuable lie a lot closer to us. We'll explore the Trojans some day, but it won't be in this century.

The most intriguing objects—and the ones worthy of an early closeup look—are the Earth Approach Objects, or EAOs. Asteroid expert Dr. George Wetherill has persuasively put forward the rationale for *in situ* examination of EAOs in a recent paper in *Icarus*:

> ... It is becoming increasingly clear that an understanding of the Earth-approaching interplanetary objects is central to major problems in planetary science. It is almost certain that these bodies are at present the principal cause of 1- to 100-km-diameter craters on the Moon and the terrestrial planets with the possible exception of Mars. It is very probable that this has been the case for the last 3×10^9 years or more.... A large proportion of the meteorites (may be) recent fragments of these bodies.... There are good reasons to believe that at least many of these objects represent the nonvolatile portions of the nucleus of 'extinct' comets.... Certain of these bodies have been identified as recently accessible sources of raw materials for space construction and manufacturing.

Earth Approach Objects can be divided into three main groups: Aten objects, Apollo objects, and Amor objects. We can also speak of Mars-crossers and Mars-grazers.

Aten objects have orbits whose semimajor axes are less than 1 AU and which overlap Earth's orbit at their aphelia. A hundred such objects likely exist, according to EAO expert Dr. Eleanor Helin, we know of three—(2062) Aten, (2100) Ra-Shalom and (2340) Hathor.

Apollo objects have semimajor axes equal to or more than 1 AU and perihelia equal to or less than 1.017 AU (Earth's aphelion distance). Apollo orbits overlap Earth's at their perihelia. We know of 23 Apollo objects; (1566) Icarus and (1862) Apollo are the best-known. From 400 to 1,000 Apollo objects may be out there, whizzing past.

Amor objects have perihelia greater than 1.017 AU and less than 1.3 AU. Of the 20 or so known Amor objects, (1943) Anteros is the most interesting in terms of robot exploration. Likely there are 1,000 to 2,000 Amor objects circling the sun between Earth and Mars.

The names "Mars-crossers" and "Mars-grazers" are self-explanatory (Figure 1). Helin estimates about 5,000 to 15,000 such objects may exist. Table 1 lists more data.

EAOs may not be "real" asteroids; as Wetherill said, they could well be the dead husks of comets. The only way to really confirm that is to (1) do a closeup examination of a couple of live comets, and (2) do the same for some EAOs. That's one reason for rendezvous and sample-return missions to these objects; the possible enormous economic benefits accruing from mining them, should they turn out to be the gold mines some believe they are, is another. EAOs may be the key to understanding the origin of the solar system; and they may represent the greatest threat to the human species, and much other life on this planet, since the invention of nuclear weapons.

Helin and Eugene Shoemaker have estimated the collision rate for EAOs with Earth as being on the order of from three to four every million years. EAOs are quite small—the largest known, the still-unnamed 1978 SB, is around 10.4 km in diameter with an estimated volume of 590 km^3. The others are all smaller, with (2062) Aten probably less than a kilometer in diameter.

What would happen if a 10-kilometer-diameter asteroid, or comet head, hit Earth? The first description of the consequences of such a strike from outer space appeared in the pages of this magazine 16 years ago, in a terrifying article called "Giant Meteor Impact." (J. E. Enever, *Analog*, March 1966.) The real thing might have happened 65 million years ago; several scientists say they've found evidence an asteroid impact wiped out the dinosaurs and a large fraction of other Earth life. Larry Niven and Jerry Pournelle described the results of such an event on human civilization in their novel *Lucifer's Hammer*.

Suffice it to say we wouldn't like it to happen, and we'd like to be able to know ahead of time if it were going to; a few well-placed nuclear explosions several years ahead of impact would be enough to change the object's orbit to one safer for us. The more we know about the existence, numbers, orbits, and compositions of EAOs, the safer we are. It's as simple as that.

Most of the same reasons for learning more about EAOs apply to the asteroids of the inner main belt. High in metals and silicaceous compounds, they may not initially be economically competitive with lunar or Earth-approach mining operations, but they're worth looking at anyway.

It never pays to have all your valuable eggs in one basket. The United States has that problem with its supply of oil, chrome, and cobalt. The human race has

TABLE 1. Data on selected asteroids.*

Name	Type	Diam. (km)**	q(AU)	Q(AU)	A(AU)	e	I(°)	Period (yrs)	Discovered
1 Ceres	MB/C	1025.0	2.55	2.98	2.77	.076	10.598	4.60	1801
2 Pallas	MB/U	583.0	2.11	3.42	2.77	.232	34.880	4.61	1802
3 Juno	MB/S	249.0	1.98	3.35	2.67	.255	13.002	4.36	1804
4 Vesta	MB/U	555.0	2.15	2.57	2.36	.089	7.144	3.63	1807
433 Eros	EAO/C	20.0	1.13	1.78	1.46	.223	10.828	1.76	1898
1566 Icarus	EAO/U	1.9	0.19	1.97	1.08	.826	22.945	1.12	1949
1943 Anteros	EAO/?	?	1.06	1.80	1.43	.256	8.700	1.71	1973
2062 Aten	EAO/S	1.1	0.79	1.14	0.97	.182	18.935	0.95	1976

*Type: MB = Main Belt, EAO = Earth Approach Object, C = Carbonaceous, S = Silicaceous/Stony, U = Unclassifiable, ? = not classified; q = perihelion; Q = aphelion; A = semimajor axis; e = eccentricity; I = orbital inclination. Data from following sources: Bender, Bowell, Pilcher, TRIAD files in *Asteroids*, T. Gehrels, ed. University of Arizona Pr., 1979; "Table 3. Minor Planets," *Anchor Dictionary of Astronomy*, V. Illingworth, ed., Doubleday, 1979; J. Niehoff, "Round-trip mission requirements for asteroids 1976 AA and 1973 EC," *Icarus* 31, 430-438 (1977).

**Diameters for asteroids 1-4 vary, even in the TRIAD files. E.g., 1 Ceres diameter given in two different places as 1016 km and 1025 km. I've used Bowell's figures.

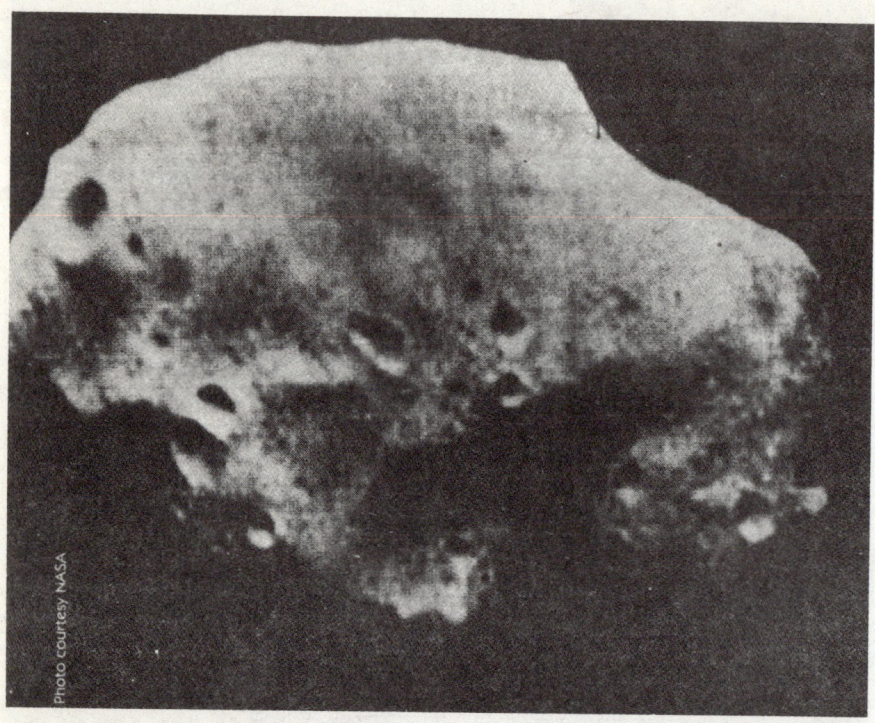

FIGURE 1. Is this what an Earth Approach Object would look like from an approaching probe? This photo of Phobos, Mars's larger moon, was taken by Mariner 9 from a distance of 5,540 km. Many astronomers believe Phobos is a large Mars-crosser captured by the Red Planet.

the same problem—one planet is not enough when the universe throws big rocks at you. It also makes no sense to depend solely on lunar mining for extraterrestrial sources of metals and material for space construction. The moon is big, close, and an easy target for a bomb aboard a "runaway" shuttle. The inner belt is none of these. Further, once the pipeline from the inner belt to Earth orbit is open, material can be delivered as swiftly as from the moon. The first load might take several years to arrive, but subsequent shipments could easily be at intervals of weeks or days.

Earth-approach objects include some S-type bodies, but most seem to be C-type, carbonaceous. They're just as valuable in their own way to an extraterrestrial construction economy. The volatiles (especially water and organic compounds) and silicaceous material are essential. So, too, are the metals from the inner belt.

It's clear: we must eventually take the measure of both inner belt asteroids and Earth-approach objects, both for the money and for the peace of mind.

GETTING THERE

Picking an asteroid to explore is not a matter of drawing a name out of a hat. Actually exploring the asteroid you do pick is also not a simple task.

First, we're looking for asteroids with orbits that have low inclinations and low eccentricities. Main belt objects average out at 9.7 degrees inclination and 0.15 eccentricity. The numbers are worse for EAOs. One Apollo asteroid, the famous (1566) Icarus, has an inclination of 22.9 degrees to the ecliptic and an eccentricity of .826. (So do some main belters: (2) Pallas has an inclination of almost 35 degrees!)

High inclinations and/or eccentricities mean higher delta-v's for spacecraft ("delta-v" is the term used to mean the velocity change made by a spacecraft when it changes its course or trajectory), and that means more fuel. That, in turn, means either a bigger probe or one with less instrumentation. Conversely, a low delta-v means a cheaper trip with more science payload for the same amount of fuel. "Cheap" is going to be the watchword for planetary exploration for some time to come, like it or not.

It would be fun, and scientifically valuable, to send a probe to Icarus. It came within 0.04 AU of Earth in 1968 (a close call if there ever was one!) and has a perihelion *inside the orbit of Mercury*. It must be a fascinating object—but the delta-v requirements for a probe to Icarus are huge, 15 to 30 kps. Icarus will have to wait.

A lot more promising is the famous Amor object (433) Eros. Even with an inclination of 10.8 degrees and an eccentricity of .223, its delta-v to rendezvous is just over 9 kps—not bad. Aten object (2340) Hathor has an inclination of only 6.27 degrees; that's good. Its eccentricity of 0.424 is not so good. (2062) Aten has low eccentricity (0.237) and a high inclination (18.9 degrees). Amor object (1943) Anteros is perhaps the best of all, with an orbital eccentricity of about 0.26 and an inclination of only nine degrees to the ecliptic. Table 2 lists other good candidates.

TABLE 2. Delta-v figures for selected asteroid missions.

Asteroid	Orbital Type	Launch Date	Mission	Post-Launch Delta-v To Rendezvous
Watts	Main Belt	6/18/86	Flyby	1.4 kps
Gagarin	Main Belt			
Vesta	Main Belt			
Mathilde	Main Belt			
Spirea	Main Belt			
Werdandi	Main Belt	6/3/88	Flyby	1.7 kps
Ceres	Main Belt			
1924/69BA	Main Belt			
Aten	Aten	7/13/92	Rendezvous	3.2 kps
Anteros	Amor	8/20/93	Sample Return	2.7 kps

Based on data from Refs. 8-11.

Besides having a low eccentricity and orbital inclination, the ideal asteroid target should be close enough to Earth to provide a reasonable trip time. Remember, we're not exploring asteroids *only* for the science; there's a lot of money and perhaps the future of the human race riding on these robot examinations. Six years may be fine for a Pioneer 11 trip from Earth to Saturn via Jupiter. That may be a little long for governments and corporations planning for profits.

(Voyager 1 took only three years to get to Saturn; that, however, was a flyby. A rendezvous-and-probe, or a sample-return from Titan, would take longer and/or lots more energy.)

Most of the asteroids we want to take a look at are close enough to make one-year round trips feasible. More likely, they'll run from two to five years, and five years is probably the upper time limit for the missions we want to run.

The Trojan asteroids look intriguing; what *is* that stuff that covers their surface? The problem is they're too far away, in Jupiter's orbit, to make a sample-return mission feasible any time soon. On the other hand, rendezvous and sample-return missions to asteroids like (2062) Aten and (1943) Anteros come in time periods from one to three years. It's even possible to do multiple-flyby missions to main belt asteroids in periods of from two to three years.

Once we get there, what do we do? In most cases we do not just fly by, taking pictures *à la* Mariner 4 or Voyager. True, we can gather enormous quantities of data that way, both with cameras and other remote sensing instruments. What we really want, however, is a "geologist's hammer" approach to asteroid exploration: rendezvous with penetrators or landers, or a sample-return.

So here are some instruments our asteroid probes should carry.

ORBITER: (1) special electronic cameras called CCDs (Charge-Coupled Devices), which are TV cameras that use a target screen of thousands of special semiconductors instead of a normal vidicon tube to detect incoming electromagnetic radiation (CCD cameras are much more sensitive than normal TV cameras; these would have a number of different filters and wide- and narrow-angle lenses); (2) X-ray and gamma-ray fluorescence instruments, which will provide definitive chemical analyses of major element abundances and particularly important minor or trace elements—hydrogen, oxygen, carbon, aluminum, silicon, titanium, iron, uranium, and thorium; (3) a mapping spectrometer; (4) a multispectral radiometer; (5) a radar altimeter; (6) fields and particles package; (7) micrometeoroid detector.

LANDER: (1) a multispectral camera for closeup imaging; (2) an alpha-proton X-ray fluorescence instrument for total chemical analysis of the surface, including all minor elements; (3) seismometer; (4) magnetometer.

PENETRATORS: If used instead of or in combination with landers, each would carry instruments to measure seismic activity, heatflow, and subsurface chemistry. The forebody, with most of the instrumentation, would be connected to the afterbody (left on the surface) by a trailing umbilical. The afterbody would contain a camera and telemetry electronics. A small Radioisotopic Thermal Generator (RTG), a form of nuclear power plant using plutonium, would be in the forebody and would provide power.

SAMPLE-RETURN: Some of the instruments carried by a lander—for example, a CCD camera and alpha-proton-X-ray fluorescence package, would be advisable on the lander/sample gatherer. In addition, it will carry a corer/scooper to obtain surface and subsurface samples totaling one kilogram in mass. Either the entire lander would be equipped to make the return journey, or a separate return stage would do it.

Those are the parameters for where we want to go and what instruments we'll carry. (Table 3.) What follows is a series of missions we could launch to the asteroids, some as soon as five years from now. The first two were originally detailed by John Niehoff of Science Applications, Inc., Schaumburg, Ill., and are

TABLE 3. Payloads for asteroid rendezvous.

Main Bus (Mass = 340 kg)	
Instrument	Mass
1500-mm CCD camera (800 × 800), multispectral (8 filters) 250-mm CCD camera (800 × 800), multispectral	20 kg
X-ray fluorescence	2 kg
Gamma-ray	17 kg
Mapping spectrometer	10 kg
Multispectral radiometer	7 kg
Radar altimeter	12 kg
Fields and particles package	10 kg
Micrometeoroid detector	3 kg
Tracking	—
TOTAL INSTRUMENT MASS	81 kg
Lander (Mass = 80 kg)	
Facsimile Camera	1.5 kg
Alpha-backscatter/proton/x-ray fluorescence	2.0 kg
Seismometer	2.0 kg
Magnetometer	0.6 kg
TOTAL INSTRUMENT MASS	6.1 kg

TOTAL MASS OF RENDEZVOUS PROBE: Main bus structure, solar panels, comdish, etc., 259 kg; + main bus instruments, 81 kg; + two landers (80 kg × 2), 160 kg = 500 kg.

missions to the EAOs Aten and Anteros. The main belt missions were proposed by Niehoff; also by a team of European scientists for ESA (European Space Agency) missions to be launched on the new—and now operational—Ariane expendable launcher.

ASTEROID MISSION: ATEN

Eleanor Helin and Eugene Shoemaker discovered (2062) Aten in 1973 during their Palomar Planet-Crossing Asteroid Survey. It's the first known asteroid with a semi-major axis less than that of Earth. The orbital characteristics and other information for Aten are listed in Table 1.

The next favorable launch window to Aten is between 1991 and 1993, giving us plenty of time to develop a spacecraft and mission plan. By the beginning of the next decade the space shuttle will be a tried and true launch vehicle, and the current problems with advanced upper stages for planetary mission vehicles will (it is hoped) be long solved.

(And if NASA doesn't solve it, ESA may do it for them, with Ariane.)

The problem with an Aten mission is the asteroid's somewhat high orbital inclination of 18.9 degrees. That means a rather high delta-v to get the probe out of the Earth's orbital plane and into Aten's, and puts some constraints on the type of mission we can fly. A straight ballistic flyby would be the easiest, but it produces not much more than a series of quick snapshots as the probe zooms past Aten. A sample-return would be great, but the delta-v for such a mission is prohibitive—easily over 7.7 kps *after* Earth-escape.

However, a rendezvous-and-lander mission *does* have an acceptable delta-v after the probe escapes Earth gravity. Delta-v following Earth-escape is just a little over 3.2 kps. That includes outbound midcourse maneuvers, rendezvous impulse, and station-keeping energy requirements. A "park-and-poke" mission is the way to go.

The Aten Rendezvous Probe, or ARP, would be launched in July of 1992 using a single shuttle, plus upper stages like IUS-twins and a spinning upper stage (or their equivalent; perhaps a souped-up Centaur or a totally new Space Tug-type upper stage).

Flight time to Aten is 198 days. ARP would arrive at Aten in late December 1992 or early January 1993, depending on the precise launch date. At no time would the probe be more than 22.5 million km from earth. Figure 2 shows a graphic of the Aten mission.

ARP would have a total weight (or more properly, *mass*) of 500 kg, about twice that of the Pioneers that went to Jupiter and Saturn, about 300 kg less than the Voyagers. The main bus would mass about 340 kg. Two landers would mass 80 kg each. If penetrators were used instead of or in combination with a lander, they'd mass about 40 kg each.

We've already mentioned the kinds of science instruments an asteroid probe

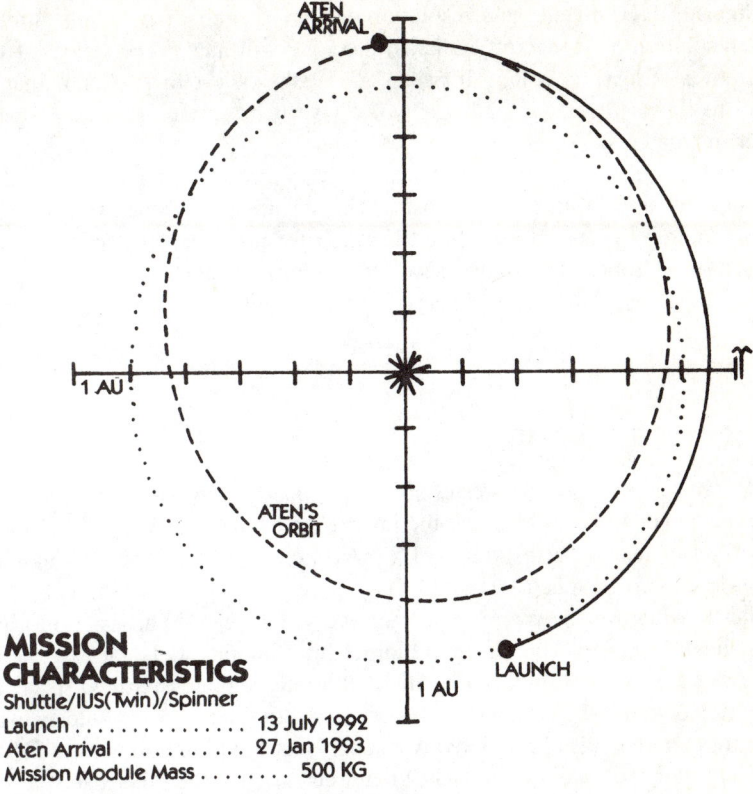

FIGURE 2. Diagram of Aten Rendezvous Mission. Dotted line is Earth's orbit; dashed line, Aten's orbit. Solid line is trajectory of ARP to Aten rendezvous at the asteroid's aphelion.
—(From Niehoff, 1978.)

would carry, and ARP would have a full complement—100 kg total, with 81 kg in the main bus and about 18 kg in the landers or penetrators.

(Two penetrators could be substituted for one lander with no real change in the mass of the ARP.)

The main bus/orbiter gets power from solar panels (about 90 watts would be needed to drive the instrumentation). The landers/penetrators get power from RTGs, radioisotopic thermoelectric generators similar to those used on Voyager. Three watts will do the job for the instrumentation.

Aten is very small, probably 0.9 km in "diameter." It has a very small gravitational field, too. Rendezvousing with Aten will not mean orbiting it (though it certainly could be done); likely it will be more like station-keeping, flying in formation with the asteroid. To image features an order of magnitude smaller than

Aten's diameter (100 meters) at a resolution of 2 arc-seconds (the *current* limiting effective resolution of spacecraft cameras) the ARP will have to be at least 1,000 km from Aten. Closer is better. It needn't orbit the object to get total imaging coverage; likely Aten rotates, and that will carry all its surface past the cameras and other instruments.

It may come as a surprise, but there is very little about the ARP and its instruments that is all that exotic. Most of the scientific instruments either have been flown before or have been studied in detail for now-cancelled missions like the Lunar Polar Orbiter. The only "new" technology would be penetrators—and those have been tested by NASA/Ames Research Center and the Department of Defense.

ASTEROID MISSION: ANTEROS

(1943) Anteros is an Amor-class asteroid more properly called a Mars-crosser; its aphelion lies beyond that of Mars, at the inner edge of the main belt (see Table 1; asteroid (7) Iris has a perihelion of 1.87 AU, close in distance—though not location—to Anteros's aphelion).

Launch windows to Anteros open up every 2.4 years. The asteroid's low orbital inclination makes it a very good target for exploration. In fact, it has the lowest delta-v for exploration missions of all currently known asteroids. Its eccentricity of 0.256 is not the best—it's 70 percent more than Aten's—and it limits the times for certain kinds of missions. A fast one-year sample-return shot is always possible, but with extremely high energy requirements. On the other hand, a low-energy sample-return mission can be carried out only once every 12 years. The next available launch window for such a mission is in 1992. And that's not all that bad. As with the Aten mission (in fact, as with *any* planetary exploration mission), we'll need the lead time. (Figure 3.)

So: the Anteros Sample Probe, or ASP, would be launched from Earth orbit in May of 1992 using a simple shuttle/twin-IUS vehicle. Its outbound coast will get it to Anteros by August 1993, 15 months later. The probe will decelerate, approach Anteros over a period of several days, and conduct a photoreconnaissance of the object.

We don't know too much about Anteros other than its orbital elements. We don't even know its diameter at this point. The cameras and other instruments aboard the ASP will glean information about the asteroid, look for a place to land, and get a sample.

After landing, the probe will stay on Anteros for 177 days or so—a long time, but necessary. The ASP must wait for Anteros to move to a point in its orbit for an optimal return trajectory to Earth. That happens on Feb. 14, 1994. The ASP plus sample will launch from Anteros and coast back to Earth entry on May 14, 1995. Earth entry speed: 12.7 kps. Total mission time: three years.

The total ASP mass, including propellants, will be about 1,700 kg, the same

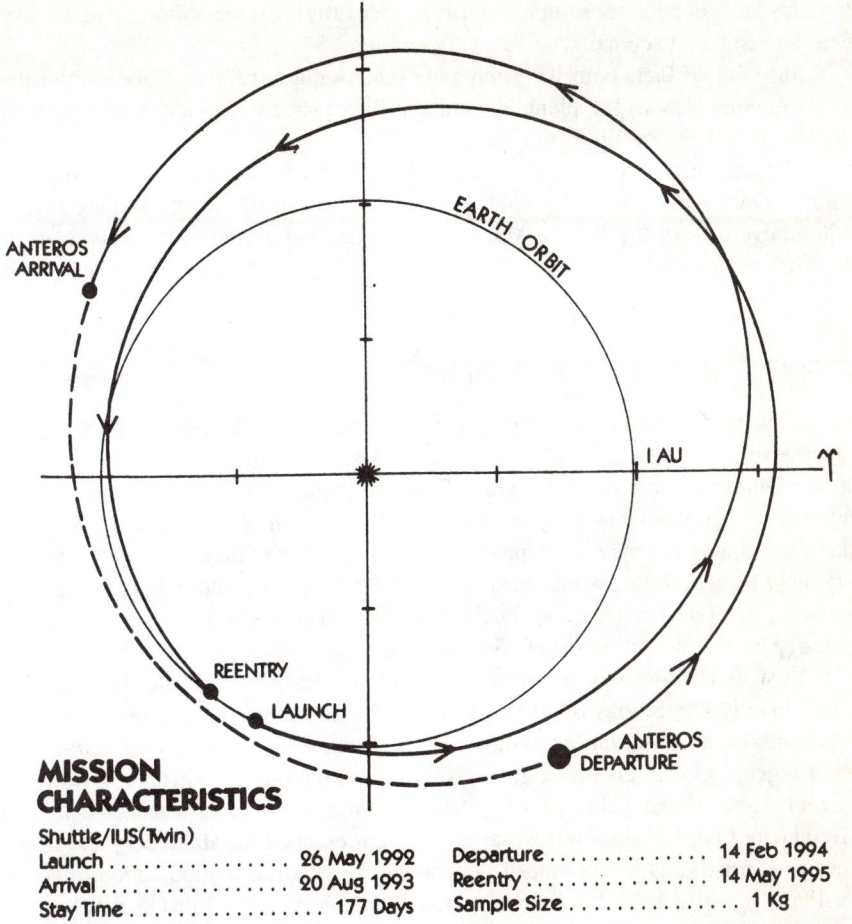

FIGURE 3. Diagram of Anteros Sample Probe Mission. Inner circle is Earth's orbit; dashed line, Anteros's orbit. Arrows show ASP trajectory to arrival at, then departure from, Anteros.
—*(From Niehoff, 1978.)*

as Voyager-plus-Earth-escape-propellants. The bus will be about 250 kg; the science payload for encounter with Anteros, around 150 kg. The lander will mass 350 kg; Earth entry capsule: 29 kg plus a 1-kg Anteros sample, for a total mass of 30 kg.

ASP's post-Earth-escape delta-v is actually less than the Aten mission's: 2.731 kps total post-escape delta-v. The biggest single increment is 1.566 kps for departure from Anteros for Earth intercept.

In one sense, the ASP is more sophisticated than anything the U.S. has ever done, since we have never carried out a sample-return mission. However, the

Russians have done lunar sample-returns successfully, and the Viking landers, with their sophisticated scoopers, were pretty complex.

In point of fact, both the Aten and Anteros missions are of about the same level of complexity as the planned Galileo orbiter/probe to Jupiter, now scheduled for launch in 1985 or 1986 (maybe . . .).

We can bring home a sample of an Earth Approach Object in eleven years. We can have a genuine two-pound piece of asteroid in our hands in time for the 19th anniversary of the Viking Mars landing—and before the third anniversary of the Aten rendezvous.

ASTEROID MISSION: THE MAIN BELT

In some ways a main belt mission is both the easiest and the hardest to do. Main belt asteroids are farther from Earth than EAOs, which means trip times can be longer. On the other hand, there are *so many* of them. That makes multiple flybys and rendezvous nearly as easy as a single-shot mission. Flybys are quicker and take less delta-v; but they don't produce a lot of good data unless the probe makes very slow passes. Rendezvous missions take a lot longer and more delta-v; but they produce a great deal of data, especially if hard lander or penetrator probes are used. It all depends on what you want. So let's take a look at examples of both.

First, what asteroids do we want to look at? At the top of almost every list of multiple flyby missions is (4) Vesta. This 555-km body is unique in the asteroid belt. Earth-based observations indicate it's covered with basalt-like material, and that suggests global chemical-geological differentiation. Is Vesta a primordial asteroid—one of the parent bodies that had time to generate internal heat and differentiate like the larger terrestrial planets, but escaped the shattering fate of the other primordials? Is Vesta indeed the source of meteorites called achondrites? Is it still geologically active? A flyby might, and a rendezvous-and-probe could, begin providing answers to the mysteries of Vesta.

Nearly as high on our list are some M-type (metallic) asteroids, like (16) Psyche. More than 200 km in diameter, Psyche seems to be a huge metal-rich object. Is it the core of some now-shattered primordial asteroid? That question could be answered, and we might also learn more about the precise chemical composition of the intriguing—and possibly valuable—M-type asteroids.

We'll also want to look at some large C-type asteroids, like (1) Ceres, the largest dark C-type body (and largest asteroid, period, at 1,025 km diameter) known.

Of course, if we get a chance we'll look at anything else interesting that comes our way, including S-types and unclassifieds. People who've studied multiple flyby missions to the main belt have found that, once you determine the flight path parameters to an initially chosen target, literally thousands of possible missions emerge.

Example one: Using an advanced Ariane booster, a 692-kg probe (space-craft mass, 395 kg) launched June 18, 1986, will do flyby encounters of (1798) Watts

(197 days post-launch), (1772) Gagarin (397 days), Vesta (479 days), and our old friend Anteros (866 days, or 2.37 years post-launch). Total post-launch delta-v: 1.4 kps. Average flyby velocity: 6.825 kps.

Another flyby example: An Ariane-launched probe massing 580 kg total (spacecraft, 290 kg, Pioneer-equivalent), leaving Earth on June 3, 1988, will encounter (253) Mathilde, (1091) Spirea, (621) Werdandi, Ceres, and asteroid 1924/69BA. First encounter occurs 197 days post-launch; Ceres encounter is 544 days post-launch. The entire mission lasts two years. Post-launch delta-v: 1.7 kps; average flyby velocity is also 6.8 kps, with the Ceres flyby at 5.7 kps, only 1.1 kps more than the Vesta flyby speed in example one.

The European scientists who drew up these (and other) scenarios believe the probe could carry about 50 kg of scientific instrumentation. They suggest a high-resolution CCD camera could get 20-meter resolution on images.

The closer we get to an asteroid and the slower we fly past it, the better resolution we get, and the better chance exists for other remote observations, such as chemical analyses with X- and gamma-ray instrumentation. However, that also means using more energy to slow the spacecraft down for a leisurely flyby. Eventually we reach the point when it's better, in terms of fuel expenditure, to forget about flybys and go to out-and-out rendezvous missions—orbits or station-keeping.

Now we're talking about something completely different from what we've discussed so far. For main belt rendezvous missions it's more fuel-efficient to forget about space-storable chemical fuels and ballistic flight paths and instead go to a probe with an ion drive—a powerplant much like the recently cancelled SEPS drive for the proposed U.S. Halley Comet rendezvous probe.

A Halley rendezvous is now out of the question for NASA; the 'lead-time' window for such a mission is long past. SEPS, though, is not completely dead and gone. Prototypes have been built, and two are in orbit, still working on experimental satellites. It is still possible to build and use a 60-kilowatt SEPS for a 1987 multi-asteroid rendezvous mission. (*Editor's Note: This article was written before 1987, which is now, obviously, behind us. But similar opportunities will exist in the future.*)

The mission would be as follows: A 500-kg mission module, with five 75-kg surface probes, would be launched on October 3, 1987 (the day before the 30th anniversary of Sputnik 1; how's that for timing!), using a shuttle/IUS-twin/60-kw ion drive. The ion drive would spiral the probe out on a non-ballistic path reaching Vesta (target Numero Uno) on May 14, 1989. The probe would station-keep at Vesta for three months, dropping a hard lander to the surface and relaying data back to Earth: chemistry, magnetism, seismic events, and lots and lots of pictures—complete global coverage of Vesta down to whatever resolution we can get away with (how close do you want to approach Vesta?).

After three months the probe departs and spirals out on its ion drive to the next target, reaching (67) Asia about two years later. Same procedure; then on to a rendezvous with (377) Campania in February 1993. Then on to Target Numero Dos: Psyche, the big metallic mass that could be the iron core of a shattered asteroid (and if it is, what a bonanza!). The probe reaches Psyche on October 7, 1994, four days before this author's 46th birthday (reasonably good timing). Yours truly gets a nice birthday present and covers the event live from Mission Control for *Analog*.

We can end the mission at Psyche; or we can send the probe on to a fifth target, (179) Klytaemnestra, with arrival on May 31, 1996.

Total mission time: 8.7 years, with asteroid encounters happening on the average of every 20 to 21 months.

AFTER THE 1990s

These missions constitute the *beginning* of our exploration of the asteroids. They will lay the groundwork for the commercial exploitation of both Earth Approach Objects and main belt objects. Mission costs will probably run between $300 to $600 million (depending on the complexity of the mission: flybys are cheapest, sample-returns the most expensive; one-target missions are more expensive overall than multiple-target missions, which return more data per buck). Those costs will, of course, not be a lump sum, but spread out over the lifetime (five to ten years) of each mission. A $600-million asteroid exploration mission, with costs spread out over ten years, would cost each American about 26 cents per year. The "crass" economic return from these missions could well be staggering.

What next? What will we do after the 1990s? Manned asteroid missions. Asteroid mining operations. The capture of a small Earth Approach Object that will be towed into Earth orbit for mining operations. Perhaps even more.

The late Dandridge M. Cole wrote about it in 1964 in his seminal book, *Islands In Space*. Take an asteroid some 30 km long and 15 km in diameter. Hollow it out (i.e., mine it from the *inside out*, going in from the ends). Cap the ends. Fill the inside with Earth atmosphere, topsoil, rivers, and lakes. Use a solar mirror to reflect sunlight down the long axis.

And then move in.

If this sounds like an O'Neill colony, you're right. If it sounds like Larry Niven's "Confinement Asteroid," or New New York in Joe Haldeman's new novel *Worlds*, right again. Great minds often move in the same direction.

And the direction is *out*.

To the asteroids.

REFERENCES

Brahic, A., et al. "Feasibility Study of a Multiple Flyby Mission of Main Belt Asteroids." *Icarus* 40, 423–433 (December 1979).

Cole, Dandridge M. "Extraterrestrial Colonies." *Navigation* 7, 83–98 (Summer-Autumn 1960).

Gaffey, Michael J., et al. "An Assessment of Near-Earth Asteroid Resources." In: *Space Resources and Space Settlements*, NASA, 1979. pp. 191–204.

Helin, Eleanor F. and Eugene M. Shoemaker. "The Palomar Planet-Crossing Asteroid Survey, 1973–1978." *Icarus* 40, 321–328 (1979).

Morrison, David and John C. Niehoff. "Future Exploration of the Asteroids." In: *Asteroids*, Tom Gehrels, editor; University of Arizona, 1979. pp. 227–250.

Niehoff, John C. "Asteroid Mission Alternatives." In: *Asteroids: An Exploration Assessment*, NASA Conference Publication 2053, NASA, 1978. pp. 225–244.

———, "Round-Trip Mission Requirements for Asteroids 1976 AA and 1973 EC." *Icarus* 31, 430–438 (1977).

O'Leary, Brian. "Asteroidal Resources for Space Manufacturing." *Acta Astronautica* 6, 1467–1480 (November 1979).

———, et al., "Retrieval of Asteroidal Materials." In: *Space Resources and Space Settlements*, NASA, 1979. pp. 173–189.

Weissman, Paul R. and George R. Wetherill. "Periodic Trojan-Type Orbits in the Earth-Sun System." *Astronomical Journal* 79, 404–412 (March 1974).

Wetherill, George W. "Steady-State Populations of Apollo-Amor Objects." *Icarus* 37, 96–112 (1979).

Stephen L. Gillett, Ph.D.

THE POSTDILUVIAN WORLD

NOVEMBER 1985

TERRAFORMING, or changing the face and atmosphere of a planet to make it more Earth-like, is an old favorite among science fiction ideas. Many writers have considered the possibility of making such a change; fewer have looked beyond the process itself to the behavior of the finished product. That, after all, is what would matter—and a terraformed planet would be just as dynamic and ever-changing as a "natural" one. Dr. Gillett did two articles for *Analog*, the first describing in some detail a possible process for terraforming Venus and the second, reprinted here, considering the behavior that might be expected of Venus *after* terraforming.

Stephen L. Gillett earned his B.S. and M.S. at Caltech, spent two years with the U.S. Geological Survey in Flagstaff, Arizona, and then went back for a Ph.D. in geology at the State University of New York at Stony Brook. His major research interests have been in applying paleomagnetism to the solution of geological problems, and in planetary sciences—a field formerly the province of astronomers, but in recent years largely taken over by "Earth scientists." After several years as an independent geological consultant in the Pacific Northwest, he recently became a research associate with the University of Nevada's MacKay School of Mines. ■

The deep gray clouds have gotten heavier, denser, and lower over the decades as the atmosphere has lightened and the temperature fallen. They begin to trail dark wisps and veils, rainfall not reaching the ground but evaporating in the still-hot lower atmosphere—as indeed often happens in deserts on Earth.

Finally, the first drops of rainwater touch the surface. They promptly flash back into steam. But they are followed by others, popping, sizzling, finally persisting on the surface briefly before evaporating. They leave a faint rime of salts behind, material dissolved and then reprecipitated.

At last boiling water gurgles along the surface for a distance. It soaks into the regolith, only to explode in steam as it encounters the hotter material beneath, just like water poured on a hot campfire. The bursting steam scatters dirt, pocks the surface, and churns the regolith.

Later, as the surface cools still more and the first tentative drops have given way to a torrent, gigantic flash floods scour the virgin dirt. Rock debris that has never known running water is channeled and transported, only to be redeposited in deltas, bars, and alluvial fans as the slopes flatten and the water slows. Mudflows flood across the surface into the lowlands, where the water gradually begins to pool. The bottom of these deepening bodies of standing water is a seething chaos, as the water soaks deeper into the still-hot material on the floor. Steam continues to rise, as from a lava flow pouring into the sea.

Salts are dissolved and carried along, to wind up in the growing ponds. The embryonic lakes and seas become brinier . . .

When the Big Rain falls on a terraformed Venus, the initial effects will be profound and spectacular. The surface of Venus has never known water action—at least not in the last 4 billion years or so. What will happen as water starts its enormously effective erosion, and the surface evolves under its new Earthlike atmosphere? Let's see.

In an earlier article ("Second Planet-Second Earth," *Analog*, December 1984), I gave some idea about how the terraforming might be done. The exact mechanism isn't important here. I make two assumptions, however: (1) the slow rotation rate remains unchanged (it would be very difficult—i.e., expensive—to change, and can probably be lived with); and (2) to avert Venus's reverting to its present uninhabitable state through the runaway-greenhouse effect, the amount of water brought in is severely limited, and the resulting small oceans are highly salty.

To backtrack for a minute: The runaway-greenhouse effect prevents liquid water from occurring on a planetary surface above a certain critical temperature. Above this temperature, the increase in the greenhouse effect—that is, the trapping of solar energy by the atmosphere—from the additional water vapor evaporated is so great that temperatures soar. Highly salty, low-volume oceans delay the onset of the runaway-greenhouse effect, because water evaporates more sluggishly from brine. (Salty oceans should arise naturally, simply because Venus will have many such compounds in its surface already.) Having much less water than Earth also helps, because it becomes more difficult to enshroud the planet in a water-vapor blanket. The vapor tends to precipitate out in high latitudes and on the night side, fast enough to keep the greenhouse effect below the critical level. We'll assume that in this way the resulting climate can be made stable.

The effects of terraforming fall into three general timescales: (1) The short-term changes: What happens right away, even before the terraforming is complete, and during the early settlement by terrestrial life forms. (2) The midterm changes, which will occur rapidly on a geological scale but very slowly on a human scale, say over a few thousand to a few million years. (3) The long-term geologic changes, over tens to hundreds of millions of years.

The first timescale is obviously the most important for the immediate settlement of Venus. The last, however, is probably the most important philosophically, because if a terraformed planet is to become a permanent second home for Earthly life, its new environment must persist over geologic time, not merely the few thousand years humans reckon as a "long time."

THE NEAR TERM: NOVA TERRA

Erosion at first will be swift. Loose dirt, never kissed by running water, will be swept up in muddy streams and carried off. Arroyos will be carved; drainages established. Regolith, soaked with water for the first time, will become mud, slumping off steep slopes, nurturing a water table elsewhere. Basins will fill, to become lakes and seas. And salts, ubiquitous in the Venusian regolith, will be leached out, to end up dissolved in the growing seas. These initial changes happen on human timescales, over hours, weeks, and years.

The changes will slow down, however, as the low areas fill up with water and the loosest dirt washes away—and as the initial rain dies away. Still, lingering slope instabilities will be a hazard. Even here on Earth, for example, instabilities (that's spelled "landslides") occur occasionally, especially in geologically active areas such as places with rapid uplift, or places which were glaciated in the Ice Age and have not yet re-equilibrated with running water.

THE LAND

A stream flowing uncertainly down a makeshift channel, a maze of ponds, potholes, and ragged islands. A misty waterfall cascades improbably off a cliff to one side, its waters landing and vanishing into new talus at the foot of the slope. Eventually, they percolate through, to gather together at the base of the debris. A hot spring gushes at another point, the fissure from which it issues already mineral-encrusted.

Farther on, the main stream splashes its way down an abrupt slope to its salty sump far below. It has already incised a shallow watercourse, and an embryonic delta is building out into the salt sea . . .

Immediately after terraforming, Venus will have a highly immature topography. "Immature" here means a topography not resulting from prolonged water erosion. What does such topography look like? Let's approach this from the other direction, by reviewing the effects of running water on the evolution of a landscape.

First of all, running water doesn't like closed depressions: lakes, ponds, any basin with no outlet. Over time, closed basins tend to be destroyed from two directions. First, they fill up with debris; flowing water carrying sediment slows down abruptly as it enters the basin, and its sediment load drops. Second, if water in the basin spills over to continue downhill, it begins to erode at the spillover point. This erosion tends to carve a canyon through the wall of the basin and thereby drain it. (Niagara Falls is eventually going to drain Lake Erie in just this way.)

Second, running water also doesn't like waterfalls, and they too are rapidly destroyed, geologically speaking. A waterfall concentrates the erosive power of the stream at one point (stand under a waterfall sometime if you're not convinced) and thus tends to destroy itself.

The net effect is that the water tries to smooth out the gradient over which it flows. No sharp breaks: no lakes, and no falls. To look at it another way, the running water tries to achieve uniform energy dissipation along its watercourse.

Of course, this evening out doesn't happen right away. Some rocks are harder than others, for example; a ridge of hard rock may sustain a waterfall long after a softer rock would be worn away. Also, in some cases eroded debris is dumped into the main stream at irregular intervals by great floods coming out of tributary streams. Such debris partially dams the stream to form rapids with pools above, and only occasionally do great floods down the main stream itself flush out the accumulated debris. (The rapids on the Colorado River in the Grand Canyon are formed in this way.) Such a stream, which has a kind of chronic unevenness, is still "juvenile"—at an early stage in its history. The surrounding land has high relief and there is lots of sediment available. But if things are allowed to go to completion its gradient will also eventually become smooth.

In addition, drainage networks become interconnected or "integrated" over geologic time. Little streams become tributary to bigger streams, which in turn flow into rivers. Lots of little independent drainages are unstable. As they expand with continued erosion, they tend to connect up.

For example, one of the characteristics of the deserts in western North America—and elsewhere—is independent drainages. Washes drain mountains to end in an isolated, enclosed depression, a dry lakebed or "playa." But even in deserts these depressions are unstable, formed geologically yesterday and gone tomorrow. The isolated drainages in the southwestern US, for example, are slowly being connected by erosion working away from the drainage along the Colorado River.

We can also expect common "braided" streams on the new Venus. This is a type of streamform which develops where (1) lots of sediment is available, (2) relatively steep slopes occur, and (3) there are big variations in water flow. The "braiding" refers to a typical streambed pattern in which more-or-less parallel channels separate and rejoin, over and over in a bewildering pattern. Such channels are separated by higher areas, or "bars," which taper on both ends and which are inundated only during maximum water flow. (The antithesis of a braided stream, in case you were wondering, is a meandering stream such as the Mississippi.)

The ephemeral streams (channels in which water flows only infrequently, after major rainstorms) in the deserts of the American West are all braided. Rivers out of steep, geologically young mountains are also typically braided; one now-spectacular example is the Toutle River draining Mt. St. Helens.

Therefore, on Venus at first there will be lots of geologically ephemeral landforms: closed depressions, some containing lakes, and waterfalls. But as the drainages slowly become integrated, these landforms will be destroyed.

This is really not all that different from a lot of immature topography on the present Earth. Since the glaciers retreated about 10,000 years ago, running water has been diligently trying to rework the glacial landscapes left behind. (Glaciers have very different erosional dynamics from liquid water.) But the abundant lakes and waterfalls in the mountains and northern part of North America, as well as the often convoluted streamcourses, show how much is still to be done.

This leads into another point: Running water does not have it all its own way. It is not the only geological agent, especially on the Earth. Other processes have dynamics of their own.

Uplift, either by slow movements of the crust itself ("tectonics") or by volcanic eruptions, raises new land to be eroded. Lava flows dam rivers and otherwise rearrange the landscape. Climatic changes cause very different processes, such as glaciation, to act at different times. In fact, between uplift and climatic change, erosion on the Earth is forever playing catch-up.

As we'll see, glaciers are not likely even on a terraformed Venus. The average temperatures will just be too high. But wind erosion may be important. On Venus, there will be high temperature gradients due to the slow rotation, as I discuss below, and that means the wind will really whistle.

Hot springs will also be abundant at first. The Venus crust near the surface will be accustomed to a surface at 450°C, not one at 50°C or so. Thus, the temperature gradient under the surface will be very steep initially, and it will take a long time for the subsurface to cool simply because planets are big and rocks don't conduct heat very well. The cooling will be hastened by ground water, however. As it percolates down, the water will absorb heat and bring it back to the surface with a hot spring. Such springs will probably be mineral-rich (after all, none of the rocks involved had ever been leached by water—much less hot water—before), and they could easily be tapped for geothermal power.

THE CLIMATE

Although it is soon after dawn, the desert is already shimmering with heat reflected off the salt flat. Even so, the clouds bearing down from the east are intimidating, tan dust billowed up in a gigantic wall over which a dark squall line can occasionally be seen.

The storm strikes suddenly, wind bearing dust like Carborundum while a stinging carpet of saltating sand grains attacks at ground level. The dust storm is abruptly followed by mud, as great warm drops of rain combine with the dust in an airborne morass. As the cloudburst continues, water collects, then starts across the surface in tentative rivulets. Suddenly, the surface is overwhelmed with a sheetflood spreading across the salt flat and dissolving the minerals there, precipitated by the drying of the water from the last storm . . .

It's gonna be hot. *Real* hot. To estimate how hot, let's first consider a simple case, a rapidly rotating planet with no atmosphere at all. Its temperature will be given by the following equation, approximately:

$$T = \sqrt[4]{\frac{S \cdot (1 - A)}{4\sigma}}$$

where T is the temperature in degrees Kelvin; S is the solar constant (i.e., the amount of energy received by the planet); A, the reflectivity or "albedo" of the planet (an albedo of O means a perfect absorber; an albedo of 1 means a perfect reflector); and the σ (Greek small letter sigma) is a constant, the "Stefan-Boltzmann" constant.

Let's see what this equation gives us. Two values we can't change: The solar constant at the orbit of Venus is about 2,560 watts per meter squared, and the Stefan-Boltzmann constant is 5.67×10^{-8} watts per square meter per degrees Kelvin to the fourth.

The albedo, however, is open to some variation. Let's assume an albedo of about 0.20, or 20%—reasonable since deserts tend to reflect pretty well, especially if they contain lots of nice white salt flats. (Mars's albedo is about 17%, for example.) The surface (fortunately) won't be nearly as dark as the Moon, which has an albedo of only 7% or so. The Moon's low albedo depends on its extraordinarily absorbent regolith, which is very finely porous right at the surface. Its absorption would be substantially degraded by contact with an atmosphere.

Now let's plug these numbers into the equation. (Taking fourth roots is easy. Just hit the square-root key twice on your calculator.) We find that the average temperature is 308°K, or about 35°C. That's 95°F, for those of you still thinking in feet'n'Fahrenheit—a nice warm spring day, in the deserts of the Southwest.

That's not too bad, you say . . . sure, the equatorial regions will be too hot, but the poles and even the temperate latitudes should be nice. Unfortunately, however, that's not all. That's the theoretical temperature for a planet with *no* atmosphere. However, the terraformed Venus will have an atmosphere, and that atmosphere will have a greenhouse effect. To get the temperature of the new Venus, therefore, we have to add a fudge factor for a greenhouse effect.

Since the terraformed Venus will have an atmosphere similar to Earth's, let's look at Earth's greenhouse effect. The total greenhouse effect on the Earth is about 35°C. The main source of the terrestrial greenhouse is H_2O, both as vapor and as clouds. (As is evident on a little thought; people are dithering about a few degrees of warming which may—or may not—result from the doubling of atmospheric CO_2. But 30-odd degrees of warming already result from the H_2O.)

Adding on 35°C of greenhouse effect to Venus obviously makes things hot indeed. However, water vapor will be much less abundant (~50% less) in the new Venus atmosphere than in Earth's atmosphere, so the greenhouse effect will be proportionately less. That's the whole point of those salty seas. At a given temperature, you just can't evaporate as much water.

Although we don't know what the cloud cover will be, as discussed below, it is not as important for the temperature because the effects from clouds tend to cancel. While they tend to raise the temperature by increasing the greenhouse effect (cloudy winter days are generally warmer than clear ones), they also tend to lower the temperature because they're white, and therefore raise the albedo.

The upshot is that the greenhouse effect on Venus will be less than on Earth. With an albedo of 20% and a residual greenhouse effect of 20°, for example, we get a mean temperature of about 55° C. That's getting pretty warm, but the poles and the higher elevations should still be OK.

However, there's one more problem . . . you will note I had assumed a *rotating* planet. But Venus rotates very slowly; the terminator, the boundary between day and night, moves a leisurely 14 km/hr along the Venus equator. A person on a bicycle could outrace the Sun on Venus. On Earth, by contrast, the terminator is scooting along at 1670 km/hr, and only the fastest jets—or spacecraft—can outrace it. If, therefore, we assume a *nonrotating* planet, and carry out these calculations, the mean temperature comes out over 100°C. That's clearly unreasonable.

Before we worry about such extreme temperatures, though, we need to consider heat transport mechanisms. Lots of effects are going to try to smooth the heat distribution. For one thing, the night side of the planet is *cold*. That's unstable; winds will carry heat from the day side to the night side. In fact, the night side will be a nice heat sink that should help keep the greenhouse from running away again.

The evaporation and precipitation of water is also a highly effective heat transfer mechanism. Evaporation of water at the hottest points absorbs energy, which is released again when the water condenses, at a point far away from (and much colder than) where it evaporated originally.

Finally, the salty oceans themselves will help even out the temperature extremes, because of the large heat capacity of water. As I argue below, the oceans are going to be warm top to bottom. An ocean warmed all the way through like that will be a nice "heat bank." Over the two-month night, the seas will warm the land around them; by day, they will absorb lots of heat in warming back up.

Still, the slow rotation is obviously gonna make things more extreme. If nothing else, the temperature change from day to night will be substantial even if the mean is around 55°C. But without detailed calculations, the extreme temperatures remain problematic. The same is true for the temperatures in places like the poles.

All in all, at this point I'll assume that the heat transport is enough to make the mean temperature near 55°C. (Anyway, if it didn't work I couldn't write this article.)

Considering heat transport leads into atmospheric circulation. What will the wind patterns look like? Where will precipitation fall? Here again, the slow rotation is the single most important effect. Earth's rapid rotation forces atmospheric and oceanic circulation to curve in the direction of rotation, an effect called the "Coriolis force." (To visualize the effect crudely, a parcel of air on the equator shares the equatorial velocity of 1670 km/sec. As it moves away from the equator, however, the velocity of the ground underneath must be less; velocity is maximum at the equator, and zero at the poles. Therefore, the parcel of air has excess velocity in the new location, and it tries to move accordingly.)

The Coriolis force on Venus, however, is less than 1% of Earth's. In fact, it's so small that the contribution from Venus's *revolution* around the Sun is comparable to that from the planet's *rotation* about its axis. At least for a first approximation, then, we can assume that Venus is not rotating at all when we look at its atmospheric circulation.

The heating by the Sun is greatest at the subsolar point, the point on the equator where the Sun is directly overhead. Here, hot air will rise and flow away in all directions. Cold air from the night side of the planet will return below the warm air. In effect, the circulation forms one giant "Hadley cell," named for the 18th-century meteorologist who first proposed such heat-driven circulation from equator to pole. (Hadley circulation occurs on the Earth, but the Coriolis force breaks it up into several cells.)

This approximate radial symmetry, with winds aloft extending out in all directions from the subsolar point, is also very different from Earth's pattern. Because of Earth's rapid rotation, circulation tends to be bilaterally symmetrical on each side of the equator. On Venus, though, not only will warm air flow toward the poles, it will also flow toward the east, toward the west; toward all points on the compass. The return flow will also be roughly even around the planet, although since it is along the surface it will be more affected by topography . . . the mountains, seas, and so forth, in the way. (Note also that this cool return flow will keep temperatures more bearable near the surface.)

That rising warm air will carry a burden of water vapor evaporated from the seas. Where this water will precipitate is important for settlement, obviously, but before we get into that, let's review what causes rain.

For rain to occur, clouds must form. For clouds to form, the air must already hold all the water vapor it can, so that any excess condenses. The amount of water vapor that a given amount of air can hold decreases with temperature. Thus, what generally happens is that as warm, moist air rises, it eventually cools to the point where it is holding more water than can remain as vapor. Condensation then occurs and a cloud forms. In turn, if there's enough condensation, and it gets swept up together, it can fall as rain.

Venus's warm air will start out much drier than Earth's, because of the salty oceans. Therefore, the air will need to rise proportionately higher than Earth, because more cooling must happen before clouds can form. (Obviously, the cloud droplets will be pure water, not brine.)

On the average, then, clouds will be higher than terrestrial clouds. Cloud cover may also be less extensive, at least on Venus's day side. Once again, though, we're getting into details that require a computer simulation to know for sure. Cloud cover is a difficult thing to model even in climates we understand.

Rain will occur when the initially warm air gets high enough to wring out most of the water. I suspect this will happen mostly near the terminator, about halfway between the subsolar and antisolar points. Here, the heavier, returning cold air from the night side will collide with the warm air from the equator, and force it farther aloft where its moisture finally will condense. Near the poles, snow may even fall. Overall, this is similar to the effect that dominates the weather in North America: the collision of cold, dry Arctic air and warm, wet Carribean air over the plains in the mid-continent.

We can also figure that air will be cold and wrung dry by the time it reaches the antisolar point. On the Earth, very little precipitatible moisture makes it all the way to the poles. (The precipitation that *does* makes it tends to persist, however.)

There also won't be any Earth-type seasons. Venus's axis is upright, and its orbit is virtually circular to boot. (Who needs 'em, anyway? Four seasons are for the birds, as far as I'm concerned . . . I prefer places that stay warm all year.)

For those of you who have to have seasons, however, the slow rotation may be a substitute. In effect, night is "winter" and day "summer," with brief snatches of "spring" and "fall" at dawn and dusk. This is not so unheard-of: It is exactly the situation in Earth's polar regions, with their season-long night and day. In fact, it doesn't even hamper plant growth too much. Because of the long days, the short arctic summer is tremendously productive; Alaska grows some record-sized crops.

Because of the lack of axial tilt, the weather will probably be pretty predictable once things settle down. Still, Venus is not smooth like a billiard ball, and the polar mountains (Ishtar Terra in particular) will channel the airflow. Just as on Earth, too, mountains will tend to generate rainfall, because they force the airflow to go over them, such that it cools.

Finally, instabilities in the airflow will ensure that things don't get too monotonous. For one thing, Venus *does* rotate, albeit slowly; at the sunrise terminator, a warm front will be continually advancing into the cold air, whereas at the sunset terminator, cold air will be continually encroaching on warm. There should be some dandy storms.

Moreover, in the equatorial regions the alternation of storms near dawn or dusk with the searing heat of midday, and the equally ferocious cold of midnight, should lead to interesting effects. One may be salt-filled playas that get filled at daybreak, all their salts dissolving, only to evaporate again as the day progresses. Something similar to this, although not so periodic, occurs in salt pans in the Southwest today. In Death Valley, acres of jagged crystals are rearranged after every major rain.

Such dissolution/reprecipitation leads to a form of erosion called "salt riving." Salty water precipitates in cracks in rocks, and as the water evaporates the expansion of the crystallizing salts tends to widen the crack. It's analogous to "frost riving," in which the expansion of freezing water shatters rock. (Both effects destroy concrete in climates where wintertime roads are salted.)

With all those winds, dust storms and sandstorms are going to be common. Besides their obvious geologic effects, they may have significant but hard-to-model climatic effects. Dust in an atmosphere changes both its albedo and the distribution of heat within it.

Last, we should look briefly at the circulation of the sea(s). Venus's brand-new oceans, as I've said, will be shallow, very salty, and maybe not even all connected. And they will be warm all the way through.

By contrast, the Earth's present oceans are cold, just above freezing except for a very thin, warm surface layer. The contrast results from the nature of oceanic circulation on the Earth: Cold polar water, being densest, sinks and flows back along the sea floor toward the equator. It is replenished by warm surface water flowing away from the equator.

The global ocean circulation on Venus will be completely different. Because there will be no cold poles, the sinking of frigid polar water will not drive the circulation. Instead, the *hottest* water, that at the equator around the subsolar point, will sink and flow away on the bottom, while cooler *surface* water flows in from all around to take its place. How come? Although the subsolar water has expanded from the heating, it has also become more saline from the intense evaporation, because the salts are left behind when water vapor escapes. The increased density from this increased salinity more than offsets the density decrease from thermal expansion. Hence, on Venus the bottom water will be hotter and more saline than the surface.

We see this sort of circulation on a small scale on Earth. In the Mediterranean, where evaporation is intense, warmer, more saline surface water sinks and flows out at depth through the Strait of Gibraltar. However, the cold Atlantic bottom water is denser than this warm saline water, so that the warm water stays above it, and in fact a current of cold bottom water flows back into the Mediterranean underneath the warm flow. Because polar water *is* denser than warm saline water, circulation driven by the sinking of warm saline water is a curiosity on the present Earth. However, at times when the Earth had no polar icecaps, its global oceanic circulation was probably driven by the sinking of warm saline water, too. (We see some evidence for this in the geologic record, which I won't get into here.)

Although . . .

Although there's no Moon, the solar tide-raising force on Venus will be about

2.6 times that on Earth. Since the Sun raises tides about half the height of lunar tides already on the Earth, tides on Venus might be even larger. But their size also depends on the depths and distributions of the seas.

One way or another, it looks as though Venus could maintain a reasonably Earthlike environment. Still, the climate of a terraformed Venus is a *very* complicated problem. There's lots of interest in modeling hypothetical Earth climates now (some of these are very controversial; most of you have probably heard of the "nuclear winter" scenario). It would really be nice if someone would do some detailed simulations on the climate of a terraformed Venus (Anyone out there got a Cray lying around?).

INTERLUDE: THE GEOLOGICAL MID-TERM

A shallow, listless sea, rimmed by saltrime shores where the great stormwaves had washed seawater up far out of its usual place, stirs sluggishly under the already blazing sun. The beaches, gleaming brilliantly in the sunlight, are calcium carbonate sand, little spheres precipitated by the bitter water and rolled around. The beaches give way to tawny desert beyond.

The water itself is thick with brine shrimp and stained by algae. The horizon is almost flat. Although the 2-month daytime is still young, and it is still cool enough for an unprotected human to wander outside, the air already holds promise of the searing heat of afternoon, when water dropped carelessly on a sunbaked rock will sizzle, and the brine sea, too salty to boil, will be a cauldron. . . .

Over time, the over-steep geothermal gradients will cool into equilibrium with the new surface temperature, and the hot springs will go away except where nearby volcanic activity keeps the heat on.

As the erosion of running water continues, the topography matures. Drainages become integrated and lakes and playas fill up. The land becomes flatter as highlands are ground away. Mineral leaching goes to completion and the seas become very salty.

Erosion, moreover, will be very effective, because most of Venus will probably be subjected to occasional cloudbursts due to the radial pattern of the weather I discussed above. Even the equatorial regions will be drenched at dawn and dusk, at least once in a while.

And therein lies a problem. The major justification for terraforming is to create an environment stable and self-sustaining—that is, not requiring ongoing human maintenance—for a significant fraction of geologic time, say a hundred million years. But, as salts and nutrients are leached toward the ocean, and the topography is leveled, the environment becomes degraded.

Moreover, climatic change, which lends the geologic history of Earth such variety, is liable to be subdued on Venus. The causes of climatic change on Earth are not completely known, but major features come from the Milankovic orbital variations (the slight, periodic changes in the Earth's orbit and spinaxis orientation), and the changing distribution of sea and land due to plate tectonics. As we'll see, Venus probably does not have plate tectonics. Its orbit also has fewer perturbations, and its axial tilt remains more uniform. (For one thing, there's no Moon to cause a precession of the equinoxes.)

Will the highlands thus be scoured down completely and all nutrients buried? Or will new uplift take place?

AND IN TEN MILLION YEARS—AND BEYOND?

The barren mountain stands alone near the salt-encrusted shore of the little sea. It is surrounded by a wild landscape of dark lava, jagged and unweathered, which speaks mutely of past eruptions and promises more to come. Mild tremors have shaken the area for weeks. The mountain is clearly listless, and a re-awakening is sure to occur soon.

Finally, stressed rock atop the mountain shatters. As the slide gathers force down the mountain slope, others join it, until the entire side of the mountain gives way in a roar of tortured rock. A gray cloud boils out of the vent, billowing upward and outward, lightning flickering through it like a failing neon tube. The lower part of the cloud races along the ground, a searing debris flow, sterilizing all before it . . . but also sowing volcanic ash, powdered rock rich with new nutrients for the depleted earth. The cloud on high continues to rise, flattening out and twisting sideways as it encounters winds in the stratosphere above. Traveling sideways more than upward now, the cloud begins to rain out its bounty of volcanic ash, far downwind of the volcano itself. . . .

Plate tectonics on the Earth keeps vital nutrient elements stirred up and available for life. Although erosion tends to carry nutrients away to the sea, where they get buried, they are not locked away completely but are eventually uplifted and restored to the biosphere. Thus, Earth's unique tectonic style has a lot to do with its habitability.

As already mentioned, Venus probably has no plate tectonics. Certainly, on the radar maps there's nothing that looks like the spreading ridges and arcuate trenches that fingerprint plate tectonics on the Earth. And the map resolution now is plenty good enough to see such things if they existed.

Perhaps this absence is not surprising. The exact mechanisms of plate tectonics on the Earth are by no means understood. However, it seems that the layer of relatively weak rock in the upper mantle, the asthenosphere, is critical; convection slowly takes place in the asthenosphere, and the more rigid lithosphere above is carried along.

One way, then, for Venus to lack plate tectonics is to have no asthenosphere.

The upper mantle where an asthenosphere should exist may be dry and rigid. A smidgen of water in solution in silicate rocks greatly lowers their melting point and also makes them more plastic at high temperatures. On Earth, a little seawater is constantly being carried down into the mantle at subduction zones. This water may perform critical lubrication by keeping the asthenosphere plastic. With no oceans, subduction on Venus could not have carried down water, and therefore the asthenosphere wasn't lubed.

Alternatively, the hotter Venusian crust may simply be too buoyant to Either way, though, the tectonics of Venus look to be very different.

Once Venus is terraformed, however, its surface will be much cooler. Will plate tectonics then begin? Probably not. Starting that initial overturn, to get the spreading-subduction system going, seems to require a big nudge. When plate tectonics began on Earth is unknown, but it seems that precursors to "plate-tectonic" type activity, involving crustal overturn between spreading centers and some sort of subduction zone, were already established early in the Precambrian, as far back as the geologic record exists. Some have even speculated that the heavy meteorite bombardment that both the Earth and Moon underwent in their first few hundred million years triggered the initial crustal overturn.

We must have some sort of tectonic overturn on Venus, however, if we're to have an environment stable over geologic time. Otherwise, no nutrients get returned to the biosphere, and the transplanted terrestrial life forms will eventually starve. Does other tectonic stirring exist?

Well, the answer may be yes. Let's first think about the forces that drive plate tectonics. Whence comes the energy to split plates and raise mountains? These processes are driven by Earth's internal heat. In turn, the heat comes from the slow decay of natural radioactive elements, primarily uranium, thorium, and the radioactive isotope of potassium, potassium-40. As this heat works its way out, some of it gets diverted into mountain building. The Earth's crust and mantle are a gigantic but very inefficient heat engine.

There are probably similar proportions of natural radioactive isotopes within Venus, too, as indeed is suggested by the Venera lander analyses and the argon-40 in the Venus atmosphere (argon-40 is one product of the radioactive decay of potassium-40). Thus, Venus should have about as much internal heat as the Earth, and it has to get out somehow. In doing so, it should stir up the crust at least somewhat.

In fact, it looks as though the heat *is* getting out now. One planetary scientist has proposed that Venus presently undergoes a process we could call "candle-wax tectonics." As sediments build up in topographically low areas, the base of the sedimentary pile eventually melts as it gets deeper. The new melt is then re-erupted, either as lava or in pyroclastic eruptions, a type of eruption in which material is suspended in volcanic gas, mostly superheated steam. Moreover, other investigators from several lines of evidence think that volcanic activity is happening right now; a major eruption may have occurred in the last ten years. Indeed, with no plate tectonics Venus will have to work harder to get rid of her internal heat just with volcanos. The planet may literally be seething with activity.

Some sort of tectonic uplift must also occur. We see the continent-like uplands Aphrodite Terra and Ishtar Terra. Such uplands may be thermal bulges, resulting from the expansion of hotter rock at depth. After terraforming, such bulges may even gradually increase, since the contrast in temperature will be greater with a cooler surface.

If volcanism is occurring, it won't shut off just 'cause the surface is terraformed. Indeed, deep crustal processes won't know the surface is different. (To a rock 100 kilometers deep, things won't have changed much at all.) And such volcanism would provide the necessary mechanism, over geologic time, to keep the nutrients stirred up for the Venusian biosphere.

NOTES ON A SELF-SUSTAINING PLANETARY ENVIRONMENT

You have probably heard of the Gaia hypothesis, the idea that terrestrial life actively maintains its environment with a vastly intricate network of biologic feedback loops. The motivation for terraforming is to set up such a global system, one that can persist over geologic time despite fluctuations from geologic and astronomic change. In fact, many planets—such as Venus—probably could support indefinitely an appropriate global ecosystem that was implanted all at once with the feedback loops already in place. Such planets could not have evolved life on their own because their environments weren't forgiving enough. (If you have only 10 million years before the oceans boil, you're not going to develop a sufficiently complex ecosystem to stave off catastrophe!)

Some might claim that it is presumptuous nonsense to speak of imposing an entire global ecosystem *ex nihilo*. With the exponentially increasing information-handling ability of human civilization, however, it certainly seems that we could establish a biosphere that works initially, and that could then fine-tune itself as it evolved.

Surely Earth's own global ecosystem must have started very simply. Earth's biosphere has constantly evolved, increasing in flexibility and complexity; Gaia has certainly not remained the same on Earth over geologic time. The environment on Earth must also have been very forgiving, biotic feedback or no, because evolution must have floundered around a lot in developing the feedback mechanisms to begin with.

And last, humans are diverse, and terrestrial life is vastly more diverse yet. On terraformed Venus, although we may perhaps find only the poles congenial, *something's* going to be able to survive in the rest. Once again, the whole point of terraforming is to give evolution another chance.

In diversification there is survival.

*Chris Peterson and
K. Eric Drexler*

NANOTECHNOLOGY

DECEMBER 1987

SUPPOSE that instead of cooking, cutting, and pasting masses of bulk matter, we had machines on the scale of molecules that could build any possible structure from its tiniest components, literally assembling it atom by atom. It may sound too fantastic for serious discussion—but that is how genes build organisms. The fact that nature does it proves that it can be done, and it may turn out that the things nature does with molecular engineering are only part of a much broader range of possibilities. If humans can learn to make molecular "general assemblers"—and there is strong reason to believe they can—the potential implications range from exhilarating to terrifying, in ways far beyond the threats and promises of any previous technology.

Christine L. Peterson studied biochemistry at M.I.T., worked in semiconductor engineering, and has consulted on electronic publishing. She is former editor of the *L5 News* and has served on the board of directors of the National Space Society. She is a cofounder of the nonprofit Foresight Institute (PO Box 61058, Dept. A, Palo Alto, CA 94306) and now edits their publications, covering such anticipated technologies as hypertext publishing, artificial intelligence, and nanotechnology.

K. Eric Drexler is a researcher concerned with emerging technologies and their consequences for the future. He has written a series of pioneering technical papers on nanotechnology and a book, *Engines of Creation* (Doubleday, 1986), which is highly recommended to readers intrigued by nanotechnology. He is currently a visiting scholar at Stanford University and president of the Foresight Institute. ■

TODAY'S technology for arranging atoms is crude—we cannot rebuild a ton of graphite into a ton of flawless diamond, nor can we rearrange a cancer cell's atomic patterns to make it healthy. Even our microtechnology slings atoms around in unruly herds. The heating, sawing, etching, and spraying techniques we use to make integrated circuit chips give us feature sizes thousands of times wider than an atom.

But we are beginning to move from this bulk technology to a molecular-level technology which will enable us to place each atom where we want it. Call this ability *nanotechnology*, because it works on the scale of a nanometer, one-billionth of a meter.

To get a feel for technology at this scale, remember that atoms are objects: spherical in shape, a few tenths of a nanometer in diameter, and able to make sturdy connections with other atoms. Recall how the properties of materials result directly from the nature of these connections—picture the long "strings" of atoms in a piece of rubber as they stretch out and then recoil; picture the rigid framework of atoms that forms hard, brittle glass. Realize that tiny machines could be built using individual atoms as working parts. Then remember that such machines already exist.

NATURAL NANOMACHINES

The nanomachines found in living things operate on much the same principles as larger machines; they just have very small parts. Lately, our abilities to exploit these existing machines—once limited to such things as making cheese and wine—have grown into the industry called genetic engineering. This suggests one path to full-fledged nanotechnology.

Genetic engineering is mostly used to make proteins. We've learned to use chemistry to synthesize DNA "instruction tapes," to use enzymes to splice them together into working genes, and to transcribe these into RNA "tapes" using more enzymes. Finally, we use an impressive piece of nature's nanomachinery: the ribosome, a programmable molecular construction machine. It reads RNA tapes and bonds amino acids together in the order the tapes specify, thereby making a specific protein molecule.

To make industrial quantities of these proteins, we use preexisting collections of nanomachinery, available everywhere, and dirt cheap—bacteria. Some industrial proteins are enzymes, nanomachines which catalyze chemical reactions. They can work fast—some can handle up to a million reactions per second.

MACHINE PARTS

Other natural protein machines include the linear motor of muscle and the rotary motor (or "proton turbine") some bacteria use to drive their propellers (yes, that's how bacteria swim). The T4 phage—a virus which attacks bacteria—works like a spring-loaded hypodermic needle when it injects its DNA; it looks like something out of an industrial gadget catalog. Protein structures already serve as moving parts, bearings, and even motors—everything needed for building complex machinery.

The T4 phage demonstrates another ability we will use: when its component proteins are stirred together in a test tube, they stick together selectively and automatically self-assemble into complete viruses. Self-assembly is the key to building complex molecular machines from simple proteins.

The first steps toward nanotechnology are already being taken by protein chemists: they have begun to design new proteins. Only a short time ago this seemed an almost impossible task—although we can control the amino acid sequence of a protein, its function depends on how this linear sequence folds up into a three-dimensional shape, and folding has been hard to predict.

Scientists have had reason for pessimism about predicting a protein's folding pattern: they work with natural proteins, which were never designed to be predictable. Engineers, however, face a different problem: they can focus on designing proteins that *do* fold predictably. Protein engineers are starting to meet with success.

OTHER PATHS

Protein engineers aren't the only ones working on the road to nanotechnology. Research on molecular electronics (what the media calls "biochips") is under way around the globe; success will involve the ability to build things to atomic specification. Likewise, non-protein chemistry and advanced micro-manipulators may help in developing molecular machinery.

In 1959, physicist Richard Feynman touched on a parallel idea in a talk: he spoke of using small machines to build smaller machines (and these to build smaller machines, and so on). He observed that "The principles of physics, as far as I can see, do not speak against the possibility of maneuvering things atom by atom," and went on to remark: "But it is interesting that it would be, in principle, possible (I think) for a physicist to synthesize any chemical substance that the chemist writes down. Give the orders, and the physicist synthesizes it. How? Put the atoms down where the chemist says, and so you make the substance." (Feynman also said that these substance-synthesizing machines "will be really useless," because chemists will be able to make whatever they want without them!)

Today, we can see several paths to molecular-scale machinery, and we can describe how these machines can be used to make far more than just chemical substances. One path—protein engineering—is guaranteed to work and is already being funded for the sake of its short-term payoff. Regardless of the route, our destination is clear.

Driven forward by incentives ranging from commercial to medical to military, protein design and related efforts are sure to continue. Where will these efforts lead? A key goal of molecular engineers will be to build a programmable machine that will work like a ribosome bonding small molecules onto a workpiece according to instructions encoded on a molecular "tape." Like the ribosome, it will completely control the arrangement of atoms in its products. Unlike the ribosome, it won't just string amino acids together into chains: it will use a wide variety of building blocks, and put them together in a wide variety of ways. Thus, it will let engineers build non-protein products—including second-generation nanomachines. Call it an "assembler." It will begin a revolution.

OBJECTIONS

Before going further, two concerns should be addressed, either of which could undermine the case for nanotechnology. First, does the uncertainty principle of quantum mechanics make nanomachines unworkable? This principle makes electron positions uncertain and hard to control on an atomic scale, but nanotechnology works with entire atoms, which have much better defined locations in molecules. Second, do the thermal vibrations of heat jostle atoms enough to make nanomachines unreliable? Engineering calculations indicate that thermal vibrations place real limits on what can be done, but the limits are tolerable. One need not check the math to be sure of these answers: to meet both of these objections one need only point to the existence proof of existing nanomachinery. It *can* be built because it already *has* been built, by nature.

Heat does cause special problems for proteins, though: too much heat causes protein machinery to denature and lose function (ever burn a finger?). Too little heat causes it to freeze. Further, proteins must be in water to work properly. The fragile, finicky nature of protein makes it difficult to work with; as soon as we can, we will use our protein machines to build rugged non-protein nanomachines.

ASSEMBLERS

First-generation assemblers (however they may be built) will soon be used to build second-generation assemblers. These will be improved devices for assembling molecular structures, resembling industrial robot arms, but built on a vastly smaller scale. More robust than protein machines, they could be designed to tolerate extreme conditions: acids, vacuum, freezing, baking. They will use as tools the same reactive groups used by chemists, but will be able to apply them precisely where needed. Since they will use atoms as building blocks, they will operate in a world full of perfect, ready-made parts.

These assemblers will be able to bond atoms into virtually any stable pattern. With them, we will be able to build almost anything the laws of physics permit to exist, almost anything we can design.

One submicron-size assembler, or even a few working together, would take ages to build a macroscopic object. Instead, we will first have them build copies of themselves, replicating to enormous numbers. Given raw materials and energy, they will multiply exponentially and then cooperate in building large objects. Using atoms from dirt and energy from sunlight, they could build immense quantities of anything, from houses to hamburgers to spaceships.

Making smaller, faster computers will be an important application of nanotechnology. Instead of spraying thousand-atom-wide ribbons of metal onto flat wafers, we will use assemblers to build three-dimensional circuits to atomic precision. The physical limits to computation are unknown today, but wherever they are, we'll reach them by using nanotechnology to build atomically perfect structures.

MECHANICAL NANOCOMPUTERS

The first nanocomputers may work mechanically, not electronically. We don't need new scientific understanding to design mechanical nanocomputers; we don't even need to use much quantum mechanics. They can be designed using Newtonian mechanics, along with known constants such as the size and mass of various atoms and the strength and stiffness of molecular bonds.

With a logic mechanism based on moving atomic-scale rods, a mechanical nanocomputer with a billion bytes of storage will fit into a cubic micron—just one-thousandth the volume of a human cell. Memory access times will be in the nanosecond range for fast memory, and in the microsecond range for tape memory (the tapes are very short, and take little time to rewind). Using the same basic design, we should be able to put the computational capacity of the human brain into less than a cubic centimeter. Electronic nanocomputers will probably be faster than equivalent mechanical designs, but may not be smaller. Radiation damage becomes a real constraint for devices built on this scale; we'll need to include repair capabilities and redundancy to get high reliability.

LIMITS

With nanotechnology we'll have greatly increased computing power and the ability to build to atomic specifications. Given the proper atoms and sunlight, we will be able to build any stable arrangement of atoms, limited only by our powers of design and by physical law.

The consequences of nanotechnology are so boggling that it is sometimes easier to focus on what it *cannot* do. It won't let us turn lead into gold, or any element into any other: that would require changing the atomic nucleus, and nanotechnology works at the level of whole atoms. Nuclei are far too small and "hard" to be altered by lugging large, fuzzy atoms back and forth.

Similarly, we won't be able to use nanomachines to build smaller machines made up of nuclei. In fact, nuclei are so mutually repellent that building anything complex with them would be impossible (at least outside a neutron star); the parts would instantly fly apart.

Nanotechnology won't let us travel faster than the speed of light, or backwards in time, or let us control gravity (except in the usual way, by moving large masses around). Nanotechnology will simply let us exploit the unchanging—but impressive—possibilities inherent in natural law. Because we and everything around us are made of atoms, we should expect striking consequences from an ability to put each one exactly where we want.

Working at the atomic level doesn't limit this technology to fabricating small objects: the existence of whales shows that nanomachines can build big. To see how molecularly engineered materials would make a difference, look at the space frontier, where strong lightweight materials play a key role.

SPACE APPLICATIONS

Consider diamond. Made from carbon atoms in a regular lattice connected by standard single bonds, diamond is one of the simplest and strongest materials. Today we can make it only in tiny bits.

Building this lattice wouldn't require a programmable assembler—a simpler enzyme-like machine could layer carbon atoms into place one by one. By laying down atoms in the right pattern, such a machine could produce strands of fine, flexible diamond fiber. They would have about fifty times the strength-to-mass ratio of aluminum and would be useful for advanced composites in planes and spacecraft.

The real potential of nanotechnology appears when we consider using it to build larger objects. Nature illustrates how to construct a large object from the molecular level up: growing animals and plants use a vascular system to transport needed construction materials and devices. We already follow this model in construction buildings, using the corridors and elevators of unfinished structures to bring in more materials. Assemblers making large objects could do likewise, constructing a scaffolding and using channels to move in the materials needed for the job.

BUILDING AN ENGINE

Picture how such a process might be used to "grow" a large rocket engine. Although the assemblers could be designed to work in vacuum or dry air, we'll find it convenient to have them work in a fluid which can carry in materials and carry away waste heat. The entire construction process could take place in a steel vat large enough to hold the finished engine. Pumps and pipes are connected to the vat, circulating various fluids as they are needed.

As we watch through a window in the vat, the process starts: a "seed"—a nanocomputer holding the plans for the finished engine—is lowered into the vat, then the first fluid is pumped in. This fluid is viscous and opaque because it is a suspension of many trillions of assembler systems, each including an on-board nanocomputer. Following their programs, these assemblers bond to the seed, building up layer upon layer, like a growing crystal.

As the assemblers bond on, the seed passes new instructions to them; they form a connected solid network which is able to pass along instructions to new members as they join. These messages tell each assembler its location in the structure and whether it should invite still more assemblers to join the structure, to extend it in a particular direction. The structure grows into a translucent engine-shaped block of assemblers.

The fluid containing the excess assemblers is then washed away by a new fluid, containing needed materials. Dissolved compounds can be broken down by the assemblers to release aluminum, carbon, oxygen, and energy. The assemblers also need room to work, so the seed now directs most of the assemblers to leave the structure. They stream out, opening up corridors and capillaries. Through a window in the vat, we see cloudy fluid billowing out of the engine-shape. Only a delicate, transparent scaffolding remains—we can barely see its outlines.

Now tiny flagelia-like appendages on the remaining assemblers help sweep fluid through the capillaries, carrying fuel and raw materials. The assemblers now have instructions provided by the seed, and materials, power, and cooling provided by the fluid. They begin work.

For engine parts needing high strength, assemblers bond carbon to form diamond. They build a dense mesh of fibers rather than a simple crystal structure, making a tough composite instead of a brittle crystal. Where heat and corrosion resistance are required, the assemblers bond aluminum and oxygen atoms to form a similar composite of sapphire. Some parts of the structure are lightened with a honeycomb of voids; others, to withstand higher stress, are nearly solid. As the assemblers finish their work, fine channels remain in even the densest parts. The assemblers make their retreat through these, pumping out and sealing up chambers as they go.

Thus, the assemblers construct the pipes, pumps, sensors, motors, and other engine parts, including any onboard computers. As each one finishes its work, it escapes and lets go of the engine, floating away into the surrounding fluid with its teammates. Finally, the fluid is drained and the finished engine hoisted out; the entire process has taken less than a day.

This engine is very different from those we build today—it has no nuts and bolts, no seams. Its surface is smooth, with an opal-like translucence that results from voids with spacings like those of the pits on a laser disk. Made of materials far better than metal, its mass is a few percent of what a modern engineer would expect.

At this point, spectators might expect the finished product to be shipped under heavy guard to an art museum, but instead it goes into routine launch service. Even under heavy use, it will last a long time. Microcracks won't spread, due to its fibrous construction. Better yet, strong materials and near-perfect fabrication will allow wide safety margins and high reliability.

The alert reader may have noticed that the description above doesn't exploit nanotechnology fully: for example, the engine could have been built by assemblers working inside an extensible "skin," building a vascular system as they went along, instead of using clumsy pumps and a vat. Worse, the finished engine didn't even contain any nanomachinery to keep it in good repair, or to reconfigure it for varying uses. Why not make the structural fibers more like muscle fibers, able to extend and contract to change the engine's shape? Why not have assemblers on hand to build entirely new parts as needed?

THE ULTIMATE SPACESUIT

Imagine a small, person-carrying, spacecraft—a spacesuit—made with nanotechnology. It could have an active structure with artificial muscle, made of diamond fiber and directed by nanocomputers. While wearing it in a g-field, you wouldn't notice the suit's mass, since the suit itself would support its own weight—and yours as well, if desired. It would have the strength of steel but still be flexible. It could act soft or hard as desired.

The suit's active structure could be programmed to amplify your movements, making you stronger, or to blunt impacts from outside. It could sense the textures of objects it touched, and transmit them to your skin. Like a plant, it could use sunlight to convert the wearer's carbon dioxide exhalations into fresh oxygen. The suit could be compact and tight-fitting, yet still have enough volume for redundant systems, capable of self-repair.

RESOURCE NEEDS

We'll need energy and raw materials to build all these atomically specified materials and products, but supplies are plentiful. Plants demonstrate that solar energy is sufficient to build molecular machines, and the Sun pours out thousands of tons of light per capita per year. Today's experimental solar cells already have higher energy conversion efficiency than plants. With nanotechnology we can probably do better.

There will be no need to chew up Mother Earth to get the atoms needed—we can use the asteroids. They can supply over a hundred million tons of material per capita, including carbon, nitrogen, oxygen, hydrogen, aluminum, nickel, iron, silicon, and platinum group metals. From these we can build anything we want, from consumer products to space settlements. Using the asteroids alone, we could build 1,000 times Earth's surface area. With strong materials made using nanotechnology, each O'Neill-style space settlement could have the area of North America.

FAST INTERSTELLAR TRAVEL

For getting around between our settlements, we can use lightsails—giant solar sails with reflectors a hundred atoms thick—which move by the pressure of sunlight and tack using gravity. But these are too slow for interstellar travel; they would take millennia to reach the stars on sunlight alone. We could use giant banks of lasers orbiting the Sun to push lightsails toward the speed of light, but getting them to stop at the other end presents a problem.

Robert Forward suggests sending a sail that can split into two sails, the first unloaded, the second carrying cargo. When the time comes to stop, we would direct the lasers so that their light bounces back off the first sail toward the front of the second one, slowing it. The first sail would speed off, and be discarded. This requires that the first sail have good optical quality. A tension structure actively controlled by nanocomputers and supporting a thin-film reflector should do the job.

Another approach would use nanotechnology more extensively. Take a one-ton lightsail made of a transparent material, able to endure intense laser light. Use the lasers to accelerate it at many *gees*; it reaches over 90 percent of the speed of light in well under a year. Once it is up to speed, onboard assemblers rebuild it into a long, thin, traveling-wave electrostatic accelerator. This forms a tube 1,000 kilometers long, with a mass of a gram per meter. The cargo, containing a cubic micron of material, forms a shell a few microns in diameter.

As the tube flashes past the target planetary system, it fires its cargo backward, leaving it almost stationary but moving toward a planet with an atmosphere. It hits the atmosphere, using it to brake, while the shell radiates entry heat to keep the cargo cool.

The tiny cargo—an assembler/nanocomputer system—immediately starts building. (All known atmospheres contain carbon, which is enough to build a grain of stuff heavy enough to fall to the surface.) These replicating assemblers build a receiver to pick up instructions beamed from home. They could be directed to build rockets and explore the system, or to build a bank of lasers to brake a passenger ship now on its way. With this system, humanity could expand its occupied sphere at near the speed of light.

A DUMB PATH TO ARTIFICIAL INTELLIGENCE

We can be sure that during this expansion, we will have artificial intelligence to help us. If we don't have genuine AI before the assembler break-through hits, we'll have it fairly soon afterwards, because nanotechnology will enable us to use a brute force method.

Back in the 1950s and early 1960s, some AI researchers tried neural simulation: using computers to physically model the brain. At the time, software approaches to the AI challenge looked more promising, and they may yet produce AI before the assembler breakthrough. If not, neural simulation will still work (it must work, if done well, because brains work). Research in artificial neural systems is coming back into fashion, with often striking results: one machine with a few hundred simulated neurons learned to read English text aloud with 95% accuracy after ten hours of learning—and it had no knowledge of English programmed in. It understood nothing, of course, but machines that understand seem sure to come.

No fundamental new insights into the nature of thought will be needed to achieve artificial intelligence via neural simulation. We can use nanotechnology to study the brain down to the molecular level. Then we could transpose the neural patterns into devices that act like neurons, but faster. If done correctly, the system will work like a brain, but faster.

A crude estimate of the speed increase is easy to make. The brain's neural response time is about one-thousandth of a second; today's experimental electronic switches are 100 million times faster. (Nanoelectronic switches should be even faster.) The human brain's neural signals travel at under 100 meters per second; electronic signals travel a million times faster. Thus, a rough guess would be that the artificial brain will work a million times faster than a human brain.

Moreover, the parts are smaller, so the whole thing will be more compact. We should be able to fit a brain-equivalent device into less than a cubic centimeter. When these shorter signal paths are added in, we find that the system should be 10 million times faster than a human brain. It would also use far more power than a brain, requiring a good water-cooling system. There's no reason not to connect a number of these brains together. John McCarthy points out that if we can put one mind-equivalent in a metal skull, we can fit 10,000 cooperating minds in a building.

Before we succeed in modeling the brain, we should be able to build powerful engines of design to do automated engineering. Already we have expert systems, computer-aided design, and programs like EURISKO that are able to evolve new rules for themselves. Eventually, engineers will be able to propose goals and then choose among solutions generated by computer.

When automated engineering is combined with brainlike devices made by assemblers, technical ideas will evolve a million times faster than today. With assemblers to construct the resulting designs, technical advance will skyrocket. Goals that seemed physically possible but a million years away will be achieved in a matter of months.

REPAIRING HUMAN CELLS

With good technical AI and assemblers, many complicated tasks will become easy. Take medicine: today physicians dump drugs into the body, cut it apart, and sew it together, but to actually heal damage, they must ultimately rely on the body's capabilities for self-repair. Nanotechnology will allow direct repair of cells, producing molecular-scale effects, like drugs, but with precise, surgical control. With a combination of technical AI and disassemblers—assemblers working in reverse, taking apart objects to analyze them—we will study the structures of all of the body's molecules and cells. Nature already demonstrates the needed abilities: access to the cell, and the recognition, disassembly, rebuilding, and reassembly of biomolecules.

Call these repair devices *cell repair machines*: as small as bacteria or viruses, they will examine, take apart, and rebuild damaged molecular structures. Complex cell repair machines will need onboard nanocomputers. These micron-size mechanical computers will be able to hold more information than the cell's DNA. They will direct simpler, smaller computers which operate the machines that manipulate biomolecules.

To visualize their relative size, bring the whole system up to a larger scale. If atoms are the size of marbles, atom-manipulating tools are like fingertips. A single repair device with a small computer is the size of a truck. A large, cubic-micron computer is thirty stories high and a football field long. The cell itself is a kilometer across: the cell repair machines will have plenty of room to maneuver.

Any physical problem—cancer, infectious disease, arteriosclerosis—could be fixed by such machines. Repairing DNA would be easy: simply compare enough strands to be sure of the correct sequence. In effect, cell repair technology will place our minds in final, complete control of our bodies.

VERY LONG LIFE SPANS

Life extension is a natural consequence: the process of aging will become reversible. We need not study the aging process and how to help the body repair itself, as we do today; instead we can study healthy young tissue and how to bring older tissue to this state. We will need to use many cell repair machines in parallel to make the overall repair process fast enough. Although death remains inevitable, these techniques will bring indefinite life extension. Moreover, this brute-force approach to medicine would presumably allow a cure even for severe, long-term, whole-body frostbite.

WHEN WILL NANOTECHNOLOGY ARRIVE?

Don't let these dramatic consequences mislead you into thinking that this revolution is far off. Cell repair machines—and their software—will be *extremely* complex, but assemblers and nanocomputers need be no more complicated than industrial robots and microcomputers. Research is under way: Japan, the U.S., Europe, and the Soviet Union are all working on protein engineering and molecular electronics.

In October 1986 the U.S. Naval Research Laboratory held the Third International Symposium on Molecular Electronics, drawing contributions from around the world, including Japan, West Germany, France, Italy, England, Australia, Argentina, Yugoslavia, and the USSR. (The authors from the last two countries failed to appear.)

The field is so new that information for laymen is scarce. Most published material is in the form of technical papers and conference proceedings on developments leading toward nanotechnology; few are about assemblers and nanotechnology itself. *Engines of Creation*, written by one of us (Drexler) and published by Doubleday, is the first book to describe nanotechnology and its consequences. Not all of these consequences are pleasant.

The dangers are clear when you consider the weapons one could build using nanotechnology. Tough, omnivorous "bacteria" could spread like pollen, consuming the biosphere in days. This has been termed the *gray goo problem*. This is unlikely to happen by accident, but it could happen by malice. There is no more reason why an industrial replicator should be any more able to survive outside a vat of special chemicals than a person is able to survive without air or water.

More likely, though, are military or police applications. Imagine programmable "bacteria" for germ warfare, or a world bugged with things much like bugs, or other, more nauseating possibilities. Accidents seem less likely than abuse, and abuse may be justified as a way to avoid accidents.

We face opportunities and dangers beyond any we have yet encountered. The global technology race makes the advance to nanotechnology inevitable; no one is in a position to cry "Halt!" with any effect. We must seek a path forward, to seize the opportunities and avoid the dangers. There is reason for hope and reason for fear, but most of all there is reason to try to understand our situation and our choices, and to try to spread that understanding as wide and as fast as we can. If we fail, our future will be short and bitter; if we succeed, our future will be rich and broad beyond past dreams.

ADDITIONAL READING

Drexler, K. Eric, *Engines of Creation*, Anchor Press/Doubleday, New York, 1986 hardcover, 1987 paper.

Part Six

BEYOND TOMORROW: THE FAR FUTURE

Thomas Donaldson

24TH CENTURY MEDICINE

SEPTEMBER 1988

NANOTECHNOLOGY, in the form of molecular cell repair machines, will very likely be the cornerstone of medicine not many centuries hence. With the kinds of capabilities opened up by such technology, many of the conditions we now consider inevitable and untreatable will be routine medical problems. And many of the techniques we now consider "futuristic medicine" will seem painfully primitive.

Thomas Donaldson has contributed regularly to *Analog* since 1984. He is a mathematician and has written several long papers and one book-length research study in mathematics. He lived for 16 years in Australia. Currently, he lives in the United States, where he is developing parallel mathematical software and algorithms for MIMD Systems, a new company. An active cryonicist since 1972, he has also written extensively on biomedical subjects for life extension and cryonics periodicals. He is a suspension member of the Alcor Life Extension Foundation.

Lest anyone think he was writing about himself, he wants it understood that he wrote the opening section of this article, on human vulnerability, several months *before* he was found to have a brain tumor (for which he is now undergoing treatment). ∎

You were only ten years old, running and jumping, full of ballgames and Masters of the Universe, when they told you that you had osteogenic sarcoma. You cried and begged your mother not to let them cut off your leg. Five interns had to hold you down when you tried to run away. When you woke up, your right leg was cut away at the knee.

We don't like to think about our own vulnerability. Every day, accidents and diseases severely injure people just like us. They happen quite unexpectedly, one day everything seems to be going well, and the next day we face utter terror. And this is not a terror of Arcturian invasions, but an ordinary terror, happening to our next door neighbor or ourselves.

This article is about medicine in the twenty-fourth century. But we need to stand back from that first, for perspective.

In the seventeenth century more than 50% of all children born never lived past age ten. Nineteenth century writers like Dickens wrote death scenes of children dying of cholera, or whooping cough, or diphtheria, or scarlet fever. . . . People died at all ages. Smallpox scarred others for life. Nobody liked this fact, but it was so common that everybody lived their lives without ever thinking about it, unless it suddenly became real, for themselves, or their children. And if so, nobody else paid attention. The invisible terror remained invisible to them, and so thankfully, they went about their own affairs.

These diseases were an unseen background to life. Wealthy families would leave town every year in the cholera season. City fathers would shut up houses on quarantine. People would not often talk about their vulnerability any more than they now talk about the color of the sky.

But the other side of vulnerability is heroism. If no injury was final and we could put everything back as good as new, heroes would become impossible. Invulnerable people can't even fall in love. Nobody writes stories about them. They have no *problem*. We are even now invulnerable to diseases which once brought terror to whole cities.

To understand medicine of the twenty-fourth century we have to understand the thin line between vulnerability and invulnerability. This line dominates our lives night and day. It is so important that we learn to walk along it without ever thinking about it, any more than we think about breathing. We construct our happiness or despair on that line.

The most important point about twenty-fourth century medicine is this: that line will be *elsewhere*.

For science fiction writers, vulnerability and invulnerability are also essential. If unshielded high-dose radiation is no longer dangerous, that's bound to affect the *story*. If bad guys once "killed" can come back, or people can recover from gaping stab wounds to the head, the entire meaning of a fight is transformed. Amazingly, events in many science fiction stories depend on injuries or diseases likely to be trivial problems a few centuries from now. Who would write a story today about a little girl's death from scarlet fever? Nobody dies of scarlet fever any more.

WHAT IS HUMAN?

Of course, we might totally redesign ourselves. But some human traits are so fundamental to almost everyone's concept of who they are that they will actively resist any redesign, for themselves or for their children. In some fundamental ways people will *not* change. In other ways they may change profoundly, but the points on which they stay the same are just as important. Below I list some *essential human traits* and others where redesign might happen.

CONSTANCY AND CHANGE

Biological control implies the ability to change human design. Here are some things about ourselves we will want to keep and others we will change.

Constants:

- *External anatomy.* Our current body form and size remain important in relating to other human beings (an "attractive" man or woman). We use tools, detachable parts, to overcome defects in our anatomy.
- *Facial expressions.* We communicate with our faces.
- *Individuality.* People resist becoming hive creatures or total separation from others. Finding a balance will still give us problems.
- *Sexuality.* Sex will still give us problems. Couples will still pair and have children.

Changes:

- *Biochemistry and metabolism.* We may use new enzymes and cell constituents containing metals (gold, arsenic) and other rare chemicals.
- *Internal anatomy.* New internal organs for life in space, defense against biological invasion, renewal of worn-out parts.
- *Immortality.* Redesign so that aging does not happen.
- *Sleep.* Brain and body reorganization so that sleep is unnecessary. We will become tired far less easily. We will have more energy and not *want* to sleep.
- *Brain.* Improved sensory discriminations, broader and longer attention span. We don't currently know how our brains work. Potential improvements are hard to see but must certainly exist.
- *Sexuality.* Pregnancy may shorten or disappear entirely. Cloning may become important, but people will see clones as continuations, not as new people ("He lost all his memories and had to relearn everything").

We can imagine ways to turn people into docile slaves, to rework them as parts of spaceships, or turn them into six legged creatures living on the floor of ammonia seas. But this article isn't about what can be done to one group of people by another. That is assault rather than medicine.

We do not want anything done to us which will make us something else entirely, that will make us say: that is not me. But who are you? Here is a test for biological changes to human beings: if people can be found who genuinely want such a change for themselves, then it will happen, and in some sense is a subject about medicine. All other events constitute assaults. Assaults will happen, but they are the *problems* of medicine rather than its success.

What is most important about identity is that our bodies and our senses are not external tools. We *are* our bodies. We cannot put them on or take them off like clothing. Our nervous systems map our own particular body. We can't just put new eyes on like binoculars. If we had the eyes of eagles, we'd need the brains of eagles to use them and the *feelings* of eagles to respond to what we see. These feelings, the neurology underlying our senses and our body, are far more important to our sense of self than just a piece of anatomical equipment.

Medicine is a branch of technology, but not just a branch of technology. We must trust our doctors. They reach into parts of ourselves very important to us. Anyone who changes our body is also changing our soul. Some of the transformations I discuss in this article may not seem to deal with medicine at all. But all of them involve changes which are terribly important to us. We could not allow these things to happen lightly, or to be done by a random stranger. One way medicine differs from other fields is that it invades the soul.

And so we see one kind of twenty-fourth century terror: that you are one day kidnapped and integrated into a spaceship. And from this assault you will need healing, to return you to a human state. They may have to force you not to be a spaceship. Not only is your body injured, but also your soul. We will have to discuss how that healing could come about.

CAPABILITIES

We can best judge capabilities of twenty-fourth century medicine by looking at what living things can do now. If today's animals or plants can do something, then some way exists to do it completely under human control. We will make quasi-living creatures, not just of the size and complexity of viruses, but up to and including the size and complexity of redwood trees or whales.

Learning to manipulate living creatures is exactly like learning to use alien machines whose vanished owners built them for unknown purposes, and with locks to prevent unauthorized use. Living creatures are independent of our wishes. Their entire design aims toward their own perpetuation and defense. We can't transplant organs because of immune systems. We can't mold them like clay because they try to retain their own forms. The creatures we build will have no such independence. They needn't even have drives to find their own food.

To some unknown degree, we will also make creatures with biochemistries quite unlike any which have yet developed in nature, or that ever could develop in nature. For all current Earth life, water is the major solvent and transport chemical. But ammonia supports biochemical reactions. Silicone fluids which freeze at $-100°C$, forms of oil with higher boiling points than water—all these might support enzyme—chemistries very broadly similar to our own. Oceans of silicone fluid just won't happen naturally. But they need not happen naturally for us to make biochemistries based on them.

The simplest forms of such machines would be single cells, like macrophages. Macrophages are part of our normal defense mechanism. These cells move about inside our tissues, destroying old and foreign cells and other debris. They slip between existing cells, accessing every part of our body. Cell membranes aren't solid walls. Artificial macrophages could pass through (not between) existing cells, reaching any body tissues. They could deliver genes or chemicals, take control of their target cells' metabolism, or even replace target cells entirely by budding off a new copy. Normal macrophages use the bloodstream for transport. Artificial macrophages could use the bloodstream, too.

Artificial macrophages can also communicate with one another. They can release diffusible chemicals to guide one another's behavior. Based on what one set found in the retina, for instance, others could carry out special modifications in the visual cortex. These devices can form an integrated repair system much larger than a single macrophage.

Some repairs need delivery of materials to a repair site much faster than the vascular system provides. We'll need devices to grow their own support tissues into a patient. For instance, severe crushing or mangling injuries require us to provide a new vascular system. The repair device might resemble a fungus, growing mycelia into the injured tissue. The mycelia would grow between existing cells rather than destroying any. We can call such devices *repair nets*. A repair net can work together with a whole family of macrophages. For instance, the macrophages could reach their target through the mycelia of the net.

Many of our organs have few provisions for self-repair. That must be done externally. If our heart is injured, we lack a backup heart. Furthermore, some modifications and transformations require external support simply because our bodies have no way to bring enough materials and energy to the repair site. The final level of repair machine actually takes over metabolism of a patient from the outside.

We can therefore expect devices which would enfold a patient completely and carry out repair. We have a model for such devices already, the womb. It supports the infant by external supply of blood and oxygen and external removal of wastes. But such a device would have even greater powers to control growth and development of the patient inside it. It would take over from the patient's own genes, controlling growth according to its own program. It would have a *brain*, to manage its control and maintain homeostasis. It could take apart a patient's entire body cell by cell, rework it into something new, and return the cells to their original location. I will call such a device a *chrysalis*.

Chrysalises, repair nets, and macrophages are *types* of machines. They would have many specialized forms for different jobs. They would have programmability. Some would be adapted to replacing only particular organs. Some repair nets could force rapid wound repair. A broken bone, for instance, would be set inside a repair net bandage. This would grow into the tissue, controlling and promoting repair. Others would be adapted to reworking and repairing nervous tissue. Some chrysalises would specialize in reviving "dead" tissue, such as severed limbs. These would be reattached after revival by another chrysalis.

The ability to design whole animals and plants to specification also means complete control over existing creatures, their metabolism, growth, and development. Since cancer is a disease of growth and development, we can expect that cancer would be a long-vanished problem. But this includes not just cancer, but also wound healing, loss of limbs and organs, and aging.

Growth and development are now almost completely out of our control. Whenever we now want to alter the shape of bone or tissue, we use surgery. But all surgery is only a crude makeshift. A medicine based on control of growth and development would treat problems very differently. We would alter shape of bone or tissue by a kind of directed growth. Millions of macrophages would enter the patient, controlling growth or breakdown of our tissues. Similar methods could modify individual cells. A genetic modifier would be a macrophage-like cell. Millions of these would enter a patient, where they would search out target cells and individually change, remove, or modify their genes.

FIRST AID

What medical problems do we have now that we'll still have in the twenty-fourth century? The major problem will have to be physical injury: broken bones, damage to internal organs or the brain, people sliced open by machinery, knifings, gunshot wounds. Machinery will still malfunction. People will be hurt, often far from places where full-scale medical help is available.

Ideally, a first aid kit is something everyone can carry with them in their wallet. What problems could a first aid kit deal with? How would it work?

The first aid kit might consist of machines the size of an aspirin tablet. They would get some or all of their materials and energy from the patient's own tissues. People going into dangerous situations might build up their resistance by vaccines of the kind I'll describe in the next section.

The Intelligent Glue. This is a tablet of small single-celled machines. To use, you stick it to the skin near the injury. The devices soak through the skin,

WHERE ARE WE NOW—MACROPHAGES

By now (1988) virologists routinely create modified viruses to place specified genes into the DNA of target cells.

Attempts to cure one disease, Lesch-Nyhan disease, are imminent. Lesch-Nyhan disease, which causes mental deficiency and uncontrollable self-mutilation, results from a single missing gene. Because Lesch-Nyhan disease is so simple, primitive genetic surgery will work. Full genetic surgery needs much more capability. Sometimes, host cells turn off inserted genes. Again, transfer viruses can put genes into the wrong cells, where they cause new kinds of pathology.

The problem is that most genes require regulation by others. A.D. Miller and others at the Salk Institute have created a virus carrying not just a gene, but its regulator genes, into host cells (A.D. Miller et al., *Science* 225 (1984) 933–8).

To keep transfer viruses from reproducing, we create modified viruses lacking genes for chemicals essential to reproduction. We grow many transfer viruses with a second virus (called the *helper virus*). The helper makes the essential chemicals. After separation from helpers, these viruses can insert their cargo of genes into cells, but not grow in them.

The macrophages I discuss are much more elaborate. They can carry much more control machinery to recognize target cells, responding only to them, or responding differently depending on cell type or cell conditions. They can still work even if the target cell isn't functioning (viruses can't do this). They can rebuild target cell machinery other than the genes. They can also transfer many more genes, up to an entire copy of the patient's genome.

Cone, R.D., R.C. Mulligan. "High Efficiency Gene Transfer into Mammalian Cells: Generation of Helper Free Recombinant Retrovirus with Broad Mammalian Host Range" *Proc Natl Acad Sci* 81 (20) (1984) 6349–53.

Eglitis, M.A., P. Kantoff, E. Gilboa, W.F. Anderson. "Gene Expression in Mice after High Efficiency Retroviral Mediated Gene Transfer" *Science* 230 (1985) 1395–8.

Miller, A.D., M.G. Rosenfeld et al. "Infectious and Selectable Retrovirus Containing an Inducible Rat Growth Hormone Minigene" *Science* 225 (1984) 933–8.

seeking out areas of broken bone and skin. Once there, they transform into the required cells for repair. If they settled around the region of a broken bone, for instance, they would form a glue to hold together the broken bone and chondrocytes to make new bone.

The Intelligent Bandage. Repairing more extensive wounds or injuries needs a *net*. The intelligent bandage sends mycelia into the patient and grows up a support network on which repair can commence. You attach it to the patient, just like the intelligent glue.

Diagnosis and Control. A lot of contemporary medicine involves recognizing what is wrong with a patient. The attention of the doctor takes over the normal function of monitoring and control performed by our own body.

First aid devices will have brains and sensors. Just like our own bodies, they will respond to low calcium in the bloodstream by releasing more. A small package attached to the neck can take over from our normal glands. It can prevent patients from going into shock, for instance.

The Intelligent IV. Often doctors must deliver drugs or blood plasma to their patient's bloodstream. This requires placing a needle into the patient's vein, a skilled task. The intelligent IV will have teeth and tongue like an animal's. It seeks out and connects to veins or arteries. It becomes a part of the patient's body. No skill is needed.

These devices can carry their own intravenous fluids, drugs, and microscopic repair machines with them. Attach them to a patient, and they will work out what is needed and inject it into the bloodstream.

The Stasis Machine. Sometimes injuries are too serious to leave a patient awake. This device puts the patient in hibernation. It sends out macrophages to modify and protect critical tissues like the brain. The macrophages redesign these tissues so that they withstand prolonged periods without glucose or oxygen. (As if their cells turn into spores.) It then shuts down the patient.

WHERE ARE WE NOW: DEVELOPMENT

As far back as the 1930s, scientists developed indirect evidence for *morphogens*. These diffused through the body of a growing animal, and their concentration at particular locations told cells there what to do. One known morphogen is a derivative of Vitamin A, *all-trans-retinoic acid*. It controls growth of chicken limbs. Morphogens for slime molds and hydra are known (C. Thaller, G. Eichele. *Nature* 327 (1987) 625–628).

One class of DNA sequences, the *homeo box*, characterizes genes controlling development in fruit flies, flour beetles, mice, and man. Specific homeo box genes are active only in specific segments of mice and fruit flies. Mark Krasnow and others at Stanford have traced interactions between homeo box genes (one gene turns on or off another) in fruit flies (M. Robertson. *Nature* 327 [1987] 556–557).

Morphogens allow very fine guidance of repair and removal of misplaced tissues by artificial macrophages or nets. These devices could also recognize cell types and even the segment to which a cell belonged. They could turn on growth and division genes in cells where these are normally off, and guide the resulting growth.

Sheard, P.; Johnson, M. *Science* 236 (1987) 851.
Utset, M.F.; Awgulewitch et al. *Science* 235 (1987) 1379–1382.

Reviving Brains after Oxygen and Blood Flow Have Ceased.

Commonly, if you aren't drugged or cooled down, after five to eight minutes without blood flow, current medicine can't revive you. *But* this isn't the whole story. Over the last 20 years, a quiet revolution has taken place. Reviving brains injured by prolonged lack of blood flow has become a major research problem.

In 1969, K.A. Hossmann and S. Sato revived electrical activity in cats' brains left without blood for an hour. Several drugs improve recovery, including naloxone, verapamil, nicardipine, gangliosides, taurine, and others.

Unfortunately, since Hossmann and Sato, clinical treatments for brain damage haven't yet materialized. The problem turns out to be more complex than neurologists hoped. Even partial success, however, means a great advance in treating strokes and brain injury.

Neurons don't disappear when deprived of blood flow. They gradually decline over four days. Protein synthesis fails in these injured neurons, and they become hyperexcitable. These two events eventually kill them.

Since brain damage matures over days rather than instantly, macrophages can take control of injured neurons, provide any protein synthesis ability they lack, and guide them back to health. Treated brains would look inflamed. Other macrophages may weaken skull bones to allow swelling.

Most citations here are for recent work on the subject:

Baskin, D., et al. *Nature* 312 (1984) 551.
Drejer, J., et al. *Journal of Neurochem* 45 (1985) 145–151.
Hossmann, K.A., Sato, S. *Science* 168 (1987) 375.
Karpiak, S.T., et al. *Stroke* 18 (1987) 184–187.
Suzuki, R., et al. *Acta Neuropath* 60 (1985) 217.
Thilmann, R., Kiessling, M. *Acta Neuropath* 71 (1986) 88–93.

THE HOSPITAL

Medicine based on growth and development will make almost no use of surgery. Many problems requiring surgery, like cancer or heart disease, simply will not occur. Yet surgery is the most important function of hospitals today. Even without surgery, though, less portable or less common equipment will belong in hospitals. Patients will have devices growing into them or enveloping them, sometimes much bigger than they are. Here is some equipment that will probably be available.

Diagnostic Machines.

These consist simply of a fine dust of macrophage spores. Patients inhale it. The machines enter the body, explore all through it, and return through the exhaled air. Another machine catches exhaled air and reads the problem from the diagnosis machines.

The Net as Substitute for the Cast.

Accidents causing major internal injuries need more than first aid. For instance, heart and lungs might become crushed and nonfunctional. We put the patient into stasis. Repair nets for such injuries would grow into a patient, reverse the stasis in their cells, and start regrowth. They provide their own substitute heart, lungs, and blood vessels and their own autonomic nervous system to keep track of their patient.

We could treat crushed limbs the same way. The same kind of morphogens causing our limbs to grow properly in the first place can shape them for repair. A repair net would grow into the limb, guided by recognition of the injured cells and a plan for how the limb should look after repair.

Hospitals involve much more than this. Right now, we take someone to the hospital for serious conditions. Someday, our serious medical problems will be trival. Serious problems for twenty-fourth century medicine will be utterly impossible.

BEYOND THE BOUNDARIES OF THE POSSIBLE

Medicine doesn't just consist of cures for known diseases which everyone believes are someday curable. It consists of cures and treatments for conditions now so far beyond treatment that they don't even have a name or seem to be diseases. Now, in 1988, we call all such conditions by only one name, "death." But "death" conceals a multitude of "diseases," each one of which must have its own individual cure. It falls apart into a million different conditions, some of which they will know how to treat, others not, and others which will be subjects of intensive research. For twenty-fourth century medicine isn't the same as twenty-sixth century medicine. . . .

The Chrysalis for Cellular Repair. Poisoning, asphyxiation, drowning—all require cell-by-cell repair, perhaps too extensive for macrophages. A chrysalis first envelops the patients, then enters in between all his cells. It disassembles the patient, surrounding each cell with its own repair machinery and vascular system. The geometry already preserves information about locations of the patient's cells. If necessary, though, morphogen chemical gradients could also retain this infor-

---

WHERE ARE WE NOW: DEATH REVERSAL

Since each form of "death" is a different pathology, I can't discuss all forms briefly. I will discuss two questions:

Survival of Personality

During an initial period after formation, chemical and other treatments can erase memories. Afterwards, however, no known treatment disrupts memories without total destruction of neurons. Permanent memories persist for years. These facts suggest (but don't prove) great durability.

Currently, the leading theory proposes that chemical changes in neurons from learning resemble those of cell differentiation. Specific genes are turned on or off when a neuron acquires a memory (cf. F. Goelet, Kandel E.R., et al. *Nature* 322 (1986) 419–422). This theory would predict that, like differentiation, memories are very durable.

Two alternatives for memory are now disproven:
Our brains don't need continued electrical activity to remember. Audrey Smith cooled hamsters down to near 0°C, stopping all electrical activity. Afterwards, they remembered mazes they had learned.

The wiring diagram of neurons to one another also doesn't code memories. Since salamanders can repair massive brain injury, Paul Pietsch cut salamander brains into pieces, scrambled the pieces, and reimplanted them. These animals recovered, and also remembered. Furthermore, molluscs such as Aplysia can achieve very rudimentary learning. Their nervous systems are small enough that we have mapped them completely. Their connections don't change with learning.

Allport, S. *Explorers of the Black Box*, W.W. Norton, New York 1986.
Goelet, P., Kandel, E.R., et al. *Nature* 322 (1986) 419–422.
Martinez, J.L., Kesner, R.P. *Learning and Memory*, Academic Press 1986.
Pietsch, P. *Shufflebrain*, Houghton Mifflin, Boston 1981.
Smith, A.U. *Biological Effects of Freezing and Supercooling*, Williams and Wilkins, Baltimore 1961.

mation. A patient would swell up to ten times original diameter. After repair, the chrysalis withdraws the same way it entered.

Since our brains contain our selves, chrysalises might work only on the brain, regrowing the rest of the body.

Chrysalises couldn't restore lost information. But information very likely survives for much longer than we can now restore life. Intensive research goes on now to revive brains deprived of oxygen for up to 60 minutes, with tantalizing successes. Beyond about an hour without O_2, metabolism of brain cells becomes a great mystery. We know that extensive structure remains on a light-microscopic scale. On electron-microscope scales, cell structures are quite disrupted. We don't know how much structure is critical or how memories are stored and how durable they will be. Our permanent memories may code by changes in activation/deactivation of genes in each neuron. This would mean that personality would survive many hours after cessation of circulation. Anthropologists have recovered fragments of brain DNA from Indians buried 3,000 years ago in a peat bog. So long as fragments of nervous tissue exist, chrysalises could reconstruct a brain around those fragments and a body around the brain.

Many current conditions, including cryonic suspension, are very simple compared to problems for which chrysalises are needed. They probably won't need more than macrophages and nets for repair.

"Death" isn't the only impossible problem. Brain and spinal injury currently devastate their victims. Active research goes on to repair these problems, too. For both rats and monkeys, normal adult neurons in their brains will divide. Control of growth and development specifically includes restarting growth in cells which have normally ceased to grow. Macrophages can enter these cells and rework the genetic controls on growth.

Ability to recover people after such serious injuries necessarily means the creation of whole classes of identity-damaged people. These will be people who have lost whole sections out of their previous lives because of some brain injury, now totally repaired. You may have been married to a woman for 40 years, but after injury and repair remember nothing of her. The way we act toward stroke patients tells me that we're certain to treat these people not as new people but as the same people they were before injury. But it will be a new kind of brain injury, one we don't have today because all such patients now die. And it will be a kind of brain injury no amount of biological control can cure.

METAMORPHOSIS

Our most important biological changes will be invisible: changes in metabolism and repair, changes in mental processing. We would use our biological control, metamorphosing into new creatures able to live in new environments. Macrophages could remodel our cells to have these traits, a kind of vaccination.

Radiation Resistance. High resistance to radiation needs active repair mechanisms for genetic and biochemical damage, much higher levels of antioxidants, and duplication of genes.

In space we also need resistance to ultraviolet radiation. Ultraviolet causes much the same kind of cell damage as gamma radiation, but to skin surfaces alone. High ultraviolet resistance may involve very high melanin production: a capacity to change skin color to deep black.

Weightlessness. Our balance perception adapts spontaneously. After Skylab, astronauts could happily sit in high speed rotating chairs without becoming dizzy. In weightlessness the major losses are of bone, muscle, and blood cells. We don't know how adaptive these changes are. Perhaps we could last indefinitely in weightlessness after initial adjustment.

But we can change ourselves to adjust quickly between gravity and weightlessness. We might have richer nets of blood vessels to bring supplies for rapid growth into muscles, bone, and marrow. Moving from weightlessness to gravity should increase production of bone, blood cells, and muscle. Reverse movement should cause equally marked loss of tissue, degradation, and storage as fat.

Decompression. The ability to live unprotected in space, even for a short time, may become extremely useful. Whales can store enough oxygen in the myoglobin in their muscle for an hour without breathing. Short periods in vacuum shouldn't cause an oxygen problem. We would need skin better able to withstand internal pressure, nictitating membranes to protect our eyes, and muscles able to close off our lungs and gut from vacuum. Since whales adapt to equal decompressions, there's no serious design problem.

Cold. Our skin darkens in sunlight. Some people will find it just as useful to grow a thick fat layer in the cold.

Not everyone would metamorphose. Protective clothing is often the best adaptation of all. But even simple adaptations to cold or vacuum become important for workers in hazardous jobs.

Metamorphosis may require such a thorough rebuilding that our metabolism cannot run normally while it goes on. With a chrysalis providing external support, quite profound rearrangements of metabolism become possible. Here's how to redesign someone to live in a cryogenic ecology on Titan. (The ecology itself may be man-created. We may create animals and plants able to live on many planets where life could not evolve naturally.)

We must rearrange someone's entire biochemistry in every single cell. First, freeze them down to the low temperature to which they are to be adapted. Second, the chrysalis grows in between their cells, much as in cellular repair. Of course, it has the proper metabolism to operate at $-200°C$. The chrysalis then individually rebuilds every cell into cells with cryogenic metabolism. It sequesters the patient's memories and maps them cell by cell onto similar structures in the target organism. The chrysalis then withdraws in reverse order. The metamorphosis is done. The man may awaken as a cryogenic creature.

Of course, metamorphosis needn't rebuild someone's whole body in all cases. For instance, our liver performs critical functions in storing energy (in glycogen) and detoxifying foreign chemicals. We may wish to rework our liver so that it will also store oxygen, as part of adapting someone to live unprotected in space. A net could substitute while rebuilding the liver from without.

IMMORTALITY

Of all suggested modifications to human beings, only one has extremely vocal, serious, and organized proponents right now. That is immortality. Its most organized proponents are the cryonics societies.

Clearly, *means* can exist to prevent and reverse aging. And they do raise an issue about identity. Most people think of themselves with finite life spans. They say: "That will happen after my time," or "I'm glad I won't be alive to see that." They have planned out their whole lives on their mortality. They go to school at a certain age, marry, have children, plan on leaving an estate on their death. Faced with the possibility, they ask if they would be themselves if they were made immortal. There is no easy answer to that question.

We can deal with major deteriorations of aging, including brain and heart breakdowns, by permanent alterations to our ability for self-repair. Modified people could go on indefinitely without growing old. Body structures like teeth, which slowly wear away, could renew themselves by slow growth of new tooth buds. If you live long enough, however, you are certain to suffer a severe accident. That accident will need external repair.

Old age has no positive evolutionary effects. Fundamentally, it happens because our former life styles rarely allowed people to live to age seventy. Now, when so many people do, we literally run out of the biological program. The symptoms of this loss of program are exactly those of old age.

Preventing and reversing old age will happen because people who did not age would have an evolutionary superiority to those who did. They would not spend so much energy supporting infants and young children. Nor would they burden their own children with their care in old age. Right now, about 50% of the population is dependent through age (too much or too little). Everyone takes this biological burden for granted, just as they once took it for granted that most children would die before age ten.

Of course, immortality rearranges a lot of common science fiction plot situations, too.

WHERE ARE WE NOW: IMMORTALITY

Intervention. Immortality (absence of aging) is a subproblem of development. Gerontologists have known that low calorie diets with all essential nutrients double life spans since work by Clive McCay in 1934. These diets cause unknown changes in hypothalamic hormones. Serotonin is somehow involved.

The major reason why aging research has not advanced further with this problem is the emotional turmoil felt by funding agencies and scientists themselves at the concept of interfering with aging.

McCay, C.M., et al. *Journal of Nutrition* 18 (1939) 1.
Timiras, P.S., Segall, P. *Federation Proceedings* 34 (1975) 83.

Evolutionary explanation. Different animals and plants live for different lengths of time. These life spans need explanation. Why don't people live as long as redwood trees?

Most evolutionary biologists would explain aging as a secondary effect. Animals in nature die of starvation, disease, or predation before they grow old enough to show old age. Their life span corresponds to just past the point at which, in nature, almost all are dead anyway. Their bodies have self-repair programs telling what to do up to that age. Our own life conditions are much better than formerly. We almost all live long enough to run out of program. Old age is the symptom of this running out of program.

G.C. Williams. "Pleiotropy, Natural Selection, and the Evolution of Senescence," *Evolution* 11 (1957) 398–411.

DISEASES WE DON'T KNOW WE HAVE

Conditions of life can change so much that traits once useful become actively harmful. It's not easy to predict conditions of life and their problems. To seventeenth century doctors, the concept that obesity could cause problems would seem far away. The problems only appear at ages few seventeenth century people would attain. Even more, obesity was then a *positive* trait, a sign of wealth and beauty. A seventeenth century doctor, once seeing the disadvantages of obesity, might still balk at doing away with this form of beauty.

We may be seeing one "disease" like this now. Getting good jobs in 1987 requires some level of technical skill. This learning takes personal traits (emotional ability to study and listen) and brain processing. Someone without these traits finds academic learning hard: this is a *learning disorder*.

Mental and emotional skills someone would need to do well in twenty-fourth century society won't just be enhancements of current abilities. To do well, we'll have to *lose* some abilities we now have. Someone from the twenty-fourth century may well appear *less* intelligent on all our current tests. For instance, current tests require testees to follow orders. With many robots available, following orders could become far less valuable.

The metamorphosis needed, a slightly rearranged brain, is simple. But our whole personality is bound up in our abilities or lack of abilities. To change these involves many questions biology won't answer. A sufficiently advanced technician might appear indistinguishable from an idiot. Do you want to become an idiot?

ILLNESSES OF THE TWENTY-FOURTH CENTURY

All currently known medical problems will be easy exercises for novice doctors. So what could go wrong? Let's look a little at history.

In the Middle Ages, wolves ranged freely over Europe. They disappeared from England first of all, but even in 1500 wolves remained abundant in France. In the winter time, they came in packs into Paris and ate children, dogs, even adults whom they found alone on the streets. And now, we no longer have problems with wolves eating our children.

But wolves are wild things. So are viruses and bacteria. In 2300 people will no more suffer from predation by wild bacteria than they now suffer predation by wolves. A second class of current diseases are *diseases of development*, like heart disease or cancer, in which normal growth and development become deranged. Fundamentally, they stem from aging, which is the most fundamental process of growth and development. Of course, all such diseases will disappear, together with aging itself.

If human beings take control of life, human beings will cause all medical problems. That is, problems will consist entirely of accidents and assaults.

Of course, crime is another form of predation. Chrysalises, nets, and macrophages could all be turned to criminal use. Specially constructed macrophages could enter our brains, change our memories and desires, and turn us into the slaves of their creator.

Accidents needn't just consist of broken bones. People can be injured by release of biological agents or toxins. Many nineteenth century doctors spent years working out causes for diseases. Even in this century, doctors have had to solve Legionnaires' disease and AIDS. Similar detective work will go on. It would involve not just seeking out harmful *chemicals*. We would have to search out harmful nanomachines, creatures able to multiply and grow. But unlike the nineteenth century or the present, such creatures would be entirely man-created.

History suggests another change, too. European populations so easily overwhelmed the peoples of the Americas because Europeans arrived carrying many deadly diseases. Centuries of plagues had given them immunity. The Indians had no such immunity. They died of terrible plagues soon after the Europeans came.

If the major disease problems of the twenty-fourth century consist of accidental or deliberate release of organisms, then we expect that people of that time will carry with them new enhancements of their immune systems. These enhancements, just like the diseases which made them necessary, will be man-created. Anyone of the twenty-fourth century will be able to resist bacterial devices which would rapidly kill any unprotected person of our time. If hostile nets can invade our brains to take control, then we'll have equal defenses against them.

Control of living things puts responsibility on us, not on the wolves. We could become very sick in the twenty-fourth century.

Nevertheless, we live at the end of a very long historical trend toward longer life spans and less sickness, and even less death from other causes. This trend tells us that people of the twenty-fourth century will live much longer lives and expect even longer ones, into millennia. But that is a statement not about medical technology but our own use of it.

NOTES AND REFERENCES

The first suggestion of a repair machine was Jerome B. White's (1969), a modified virus. The idea of macrophages and repair nets has existed at least since 1977, when Thomas Donaldson wrote a discursive bibliography ("Cryonics: A Brief Scientific Bibliography") describing means of repair, and Michael Darwin independently proposed a modified white blood cell for repair ("The Anabolocyte: a Biological Approach to Repairing Cryoinjury," *Life Extension Magazine* 1977, 80–83). In 1977 Thomas Donaldson also circulated a description of the chrysalis, though not by that name. This was published in *The Immortalist* (12 (1981) 5–10) as "How Will They Bring Us Back, 200 Years from Now."

Gordon R. Woodcock

TO THE STARS!

JUNE 1983

THERE was a time when "everyone knew" it was impossible to go even as far as the Moon. Now, a few years later, man and his machines have been there and beyond, and we are looking much farther out—thanks largely to those visionaries who refused to accept the fashionable notions of "impossible." Our future still depends on such visionaries, but there is reason to hope that the climate is a bit more favorable for them now. Going to the stars is a vastly harder job than going to the Moon, but the trip would be worthwhile—and now even serious scientists grant that it can be done. The following goes into some detail on *how* it might be done.

Gordon R. Woodcock was born and raised in Oregon and educated as an engineer, with an early interest in space flight, astronomy, and science fiction. He can remember when publishing articles on space flight was considered by some professionals as a blight on one's professional image, and he has played an important role in changing that state of affairs. Employed by Boeing for many years, he has worked on a wide variety of space-related projects. He recently played a key role in Boeing's successful space station bid and is now managing engineering studies of manned lunar and Mars missions. He has also published a textbook on space stations and platforms (Orbit Books, 1986). ■

"SPACE travel is utter bilge"—thus spake Sir Richard van der Riet Woolley, Astronomer Royal of England, in 1956.

This quotation is, in some respects, astonishing. By that date, the United States had announced intentions to launch an Earth satellite in 1957. Dozens of engineering studies of space flight had been published. There was no serious doubt as to eventual technical feasibility of flights to Earth orbit and to the moon and planets. There was only doubt as to when they might be accomplished.

But Dr. van der Riet Woolley was, after all, an astronomer. To an astronomer, then even more than now, "space" meant stars and galaxies. I don't agree with his sentiments, but if he meant that *starflight* is utter bilge, his remark is more understandable.

Astronomers are familiar with the vast, dark gulfs of deep space between the stars; they deal routinely with inconceivable distances. Although casual observers of the space travel scene may view starflight as only a modest step beyond trips to the planets in our solar system, it will in fact be inordinately more difficult.

The farthest we have ventured from Earth in manned ships is to our moon, about 400,000 km. This distance, although great compared to the distance we travel on the surface of the Earth, is insignificant compared to interstellar space. The distance to Alpha Centauri, the nearest bright star, is a little more than 40,000,000,000,000 (forty trillion) km. If we make a celestial scale model, with the distance to the moon as one centimeter (about 2/5 of an inch), the distance to Alpha Centauri will be about 1000 km (62 miles). On this scale, the distance to the sun is about 4 meters (13 feet), and the distance to Saturn, 40 meters (130 feet). At Voyager speeds it would take some 50,000 years to reach the nearest star. Skepticism is understandable.

But skepticism is almost certainly wrong. As recently as 1926, rocket flight to the moon was pronounced impossible. A futurist view more consistent with history is that, since starflight is not physically impossible, it is inevitable.

Our star-quest clearly must begin with finding a place to go. A few years ago, it was thought that places had already been found. But the extrasolar planet discoveries of the 1960s seem to have been premature, probably the result of instrument errors.

> *In light of demonstrations that the suspected perturbations were instrumental in origin, it may be said that* there is no unequivocal evidence for the existence of planets outside the solar system.—Project Orion (a NASA study of means of detecting extrasolar planets), NASA SP-436, 1980.

The question of places to go, of course, is related to processes of star formation. Again from Project Orion.

> *An equivalent way of expressing present cosmogonic hypotheses is to say that planetary system formation seems to be a natural, if not inevitable, aspect of the star formation process. The important distinction between the [old] catastrophic and [new] nebular cosmogonic hypotheses is that, if the former is correct, planetary systems are the exception rather than the rule, whereas if the latter is correct, planetary systems are the rule. A systematic study of the frequency of occurrence of planetary systems would thus provide a valuable observational check on present theories of star formation.*

Finding a place to go is thus likely to be a by-product of astrophysical research and need not be justified solely as a prelude to an interstellar flight venture. The question remaining is, "Just how difficult is it?"

There are two ways to find planets orbiting other stars. The first is astrometry: observing the very slight irregularity of motion of a star due to the gravitational effect of its planet(s). The motion to be observed is indeed slight! The irregularity in the Sun's motion, influenced by Jupiter, as seen from the nearest star, would present a perceived waviness in the motion no greater than the width of a softball in New York as observed from Los Angeles.

> *With one exception, the telescopes presently used in astrometry . . . are about 60 years old, and only two were designed for high precision astrometric observations. . . . None of the existing instruments are designed . . . for successful detection of very small perturbations.*—Project Orion.

Project Orion concluded that ground-based astrometry could be improved by more than an order of magnitude and that such improvement would make possible the detection of large (Jupiter-size) planets around stars out to about 100 light-years' distance. This volume of space takes in thousands of stars. The results of such a search would be very significant, both to the theory of star and planet formation and to questions of intelligent life in the universe.

The second way of finding planets is direct observation—that is to say, imaging—with some sort of telescope. Observation will tell us the size or color of an extrasolar planet. Astrometry, on the other hand, will tell us its mass and orbital period. The two schemes are complementary.

It is quite easy to calculate the brightness of a planet circling a distant sun. The result, for stars within ten or so light years from Earth, is on the order of 25th magnitude, a brightness about a hundred million times fainter than the faintest stars seen by the unaided eye, but easily discerned by the largest Earth-based telescopes. "So what's the problem?" one might ask. The problem is the brightness of the parent star. It will be something like a billion times brighter than its planets. Its faint planets will be entirely lost in the blazing glare of the star's image. This is why Earth-based telescopes have never seen extrasolar planets.

If we put a telescope in space, the chances of direct observation improve because the scattering effects of our Earth's atmosphere are avoided. Still, the image of a star in a telescope is not a point, but a smear of light surrounded by rings that become gradually fainter as one looks farther from the central image. At distances expected for planets, the brightness of these surrounding rings will overwhelm the faint planetary images.

Consequently, sophisticated ways to defeat the brightness of the rings have been proposed. One way is apodization: tailoring the reflectivity of the telescope's mirror to minimize the brightness of the rings. Although apodization is theoretically capable of major improvements in the sensitivity of a space telescope for faint planetary images, it requires extreme precision in telescope optics. Alternative schemes include the use of interferometry and of distant occulting disks. These seem to be more practical approaches.

In infrared light, an extrasolar planet will be much brighter relative to its parent star than in visible light, because the planet emits infrared light from its own warmth, while the visible light is merely reflected. Infrared wavelengths for best detection, however, are about ten times longer than those of visible light. Consequently, by the laws of optics, ten times greater aperture is needed for the same resolution.

An interferometer simulates a great aperture by simultaneously analyzing the images from two or more smaller apertures coupled together. An interferometer for extrasolar planet detection would spin about its boresight, aimed at the parent star. Images from the two apertures would be combined to null the bright image of the star while expressing the images of any faint planets as oscillating electronic signals at the frequency of rotation of the interferometer. Sophisticated computer analysis methods would yield very high sensitivity for this instrument.

The planetary interferometer space telescope is a rather specialized instrument, ill-suited for other purposes. The occulting disk, however, is an augmentation of an otherwise general-purpose telescope.

One occulting disk concept was proposed by this author about five years ago. The idea is to fly a disk-carrying spacecraft to a great distance from a general-purpose space telescope, in order to blot out the image of a star and make its planets more easily visible. If we select a suitable disk size and distance, the disk will not block the planet's image. Positioning of the occulting disk could be controlled by a laser beam launched from the telescope. Navigating by sensing the laser beam, the disk spacecraft could interpose itself between the star and the space telescope.

One might imagine this to be a nearly foolproof scheme, but because of the diffraction of light, it only offers an improvement factor on the order of one hundred. This, however, is enough to make extrasolar planets eminently detectable. We should even be able to detect smaller, Earthlike planets orbiting the nearer stars.

On balance, the finding of planets orbiting the nearer stars doesn't seem so difficult. Several methods have been studied, and all would apparently work. Moreover, there is a strong scientific motivation, unrelated to interstellar travel, to attempt detection of extrasolar planets.

Detection schemes, as presently understood, will be able to distinguish between gas-giant planets and Earthlike planets. They will tell us something about mass, size, color, orbital period, and temperature. But we will see no surface detail. The greatest telescopes imaginable will show us less than we can see of Mars with the naked eye.

However, a technical civilization able to accomplish interstellar travel will be able to survive in any system with Earthlike planets, whether or not they are able to harbor indigenous life. We now imagine ourselves nearly able to eke out a survival in the asteroid belt. Surely, interstellar colonists could do as well.

Places to go, according to present scientific theories of star formation, are plentiful. The progress of space research can be counted on to find a few long before we are ready to engineer a starship.

> *It is certainly out of the question, at our present level of technology or, indeed, at any level we can foresee, to mount an interstellar search [for intelligent life] by spaceship.*—The Search for Extraterrestrial Intelligence, NASA SP-419, 1977.

At any level we can foresee? Let's try some foreseeing:

First, just how tough is it? Science fiction long ago created the three themes presently known for starflight: fly slow (the generation ship), fly fast (the relativistic ship), and fly tricky (invoking some sort of spacewarp or fifth-dimensional "jump"). The third theme relies on a physics presently unknown, but the first two depend only on our ability to apply energy and power to surmount the barriers of time and distance.

We need *both* high energy and a high power-to-weight ratio to travel to even the nearest star in a "reasonable" time. High energy will give us enough rocket jet velocity to attain the speed needed to cross interstellar distances in reasonable time. Only with high power-to-weight can we accelerate to our speed in a reasonable time.

"How high is high?" is the obvious question. Ordinary rockets powered by chemical fuels have plenty of acceleration but far too little energy (specific impulse). Nuclear-powered electric rockets could achieve enough jet energy, but are limited to low acceleration by the complexity and weight of their power conversion cycles. The nuclear energy, released as heat, must be changed into electrical power to operate an ion engine. Electric rockets, however powered, confront us with the old nemesis of space propulsion engineering: power conversion cycles. Always heavy and complex, they are usually inefficient as well. I once read an article describing an electric ion engine as "The Star Engine." Maybe, if one accepts a trip time of a thousand years or more!

We have to find a way to express nuclear energy directly as jet energy. Fusion reactor concepts offer fascinating possibilities. (Figure 1.) A magnetic-containment fusion reactor (assuming we can get one of these to work) will be a magnetic "bottle" of very hot plasma. All we must do is provide a controlled leak in the bottle, so that the plasma gradually leaks out in a particular direction, producing a high-energy jet. The leaky bottle is replenished by fresh fusion fuel. Specific impulses ranging from a few thousand to as much as a million seconds have been projected by various investigators. Power-to-weight ratios as high as 10 to 100 kilowatts per kilogram might be reached, but we must first resolve a twofold problem.

The typical magnetic-containment fusion reactor concept begins with an evacuated chamber, usually doughnut-shaped, in which the reaction takes place. The

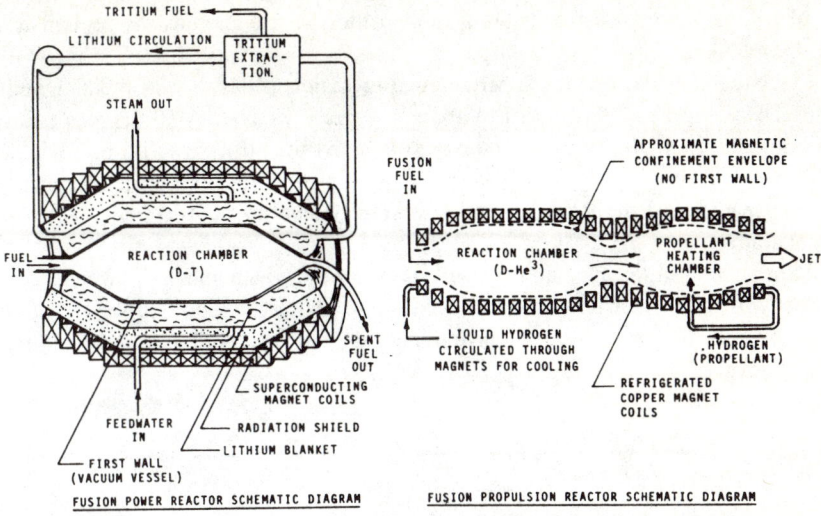

FIGURE 1. Fusion power reactor concepts for power and for propulsion are contrasted in this diagram. For simplicity, I have used a magnetic mirror geometry, although actual reactors will probably use a more complex geometry such as the Tokamak.

The power reactor includes a "first wall" vacuum vessel that maintains the high vacuum needed for the reaction; a lithium blanket in which tritium is regenerated; and superconducting magnets for confinement of the reaction. Power is extracted as heat from the lithium blanket. The heat boils water to run steam generators.

A rocket reactor may employ copper magnets cooled to cryogenic temperatures by liquid hydrogen. The hydrogen, after cooling the magnets, would be used as reaction mass for the rocket. It is assumed that this reaction runs on deuterium and helium-3. No blanket is needed. Neither is a first wall, because the natural vacuum of space will suffice.

The energy of particles leaking from the reaction may be too high, depending on the mission. If we want to trade jet velocity for higher thrust, a second magnetic bottle can be used to redistribute the energy of the reaction products to hydrogen propellant.

wall of this chamber, the "first wall," must be vigorously cooled so that it will not be vaporized by the storm of radiation emanating from the reaction. This wall is surrounded by a "lithium blanket," actually a container filled with liquid lithium metal. Part of the reaction—the fusion of deuterium and tritium atoms into helium—occurs in the evacuated doughnut. About three-fourths of the energy released comes off as very high-energy neutrons. These are absorbed in the blanket, converting its lithium into tritium and helium. The blanket is necessary because natural tritium is very scarce; tritium used inside the reactor must be replenished from the blanket.

Outside the blanket are superconducting magnets that create the magnetic bottle. These must be refrigerated to about 5 degrees K (about 450 degrees below zero Fahrenheit). If the fusion reactor is to be used for utility electric power generation (the present objective of the fusion power research program), the energy from the reactor will be taken from the lithium blanket as heat to make steam to run generators.

So we are still confronted by our space-propulsion nemesis: power cycles. The first generates electricity from the heat of the fusion reaction; the second cools the superconducting magnets. We have to get rid of these if we are to have a useful fusion star-drive.

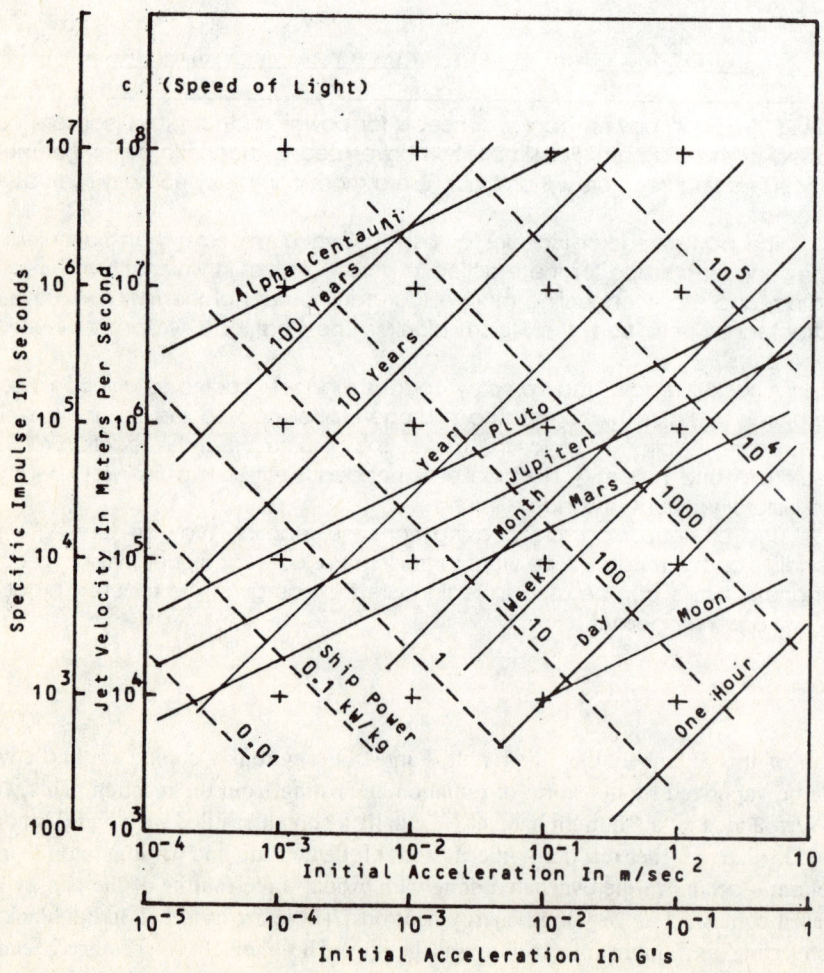

For a star-drive, we can remove the lithium blanket. This gets rid of a major heat source. We could make tritium on Earth in fusion reactors to fuel our starship. The short half-life of tritium, 12 years, is a problem for a starship whose travel time may be several times this, but the decay of tritium into helium-3 may be a blessing in disguise, as we shall see.

We should also get rid of the superconducting magnets and, as proposed by Dr. Robert Bussard, use magnets cooled by the engine's fuel flow.

We could do even better by using a different fusion reaction. There are at least three from which to choose: deuterium/helium-3, deuterium/deuterium, and boron/hydrogen, listed in order of ignition difficulty. The first of these is almost as easy to ignite as deuterium/tritium, but there is essentially *no* helium-3 on Earth. It will, however, be a by-product of a fusion power program because of the decay of tritium. Thus a fusion energy economy on Earth could make starship fuel. There is doubt, however, that it could make enough. If we began to save up helium-3 when the first fusion reactor went on-line, there still might not be enough for even the first starship. These ships will generate enormous power and have a great appetite for fusion fuel.

That is why the British Interplanetary Society, in its study of a fusion-powered star probe, proposed that helium-3 be obtained from the atmosphere of Jupiter. Unlike Earth, Jupiter still has its primordial atmosphere of hydrogen and helium;

FIGURE 2. Calculating the performance of a spaceship can be complicated. But if the ship is powerful enough, we can ignore gravity fields. It is then fairly easy. The ship will accelerate to a maximum speed and then turn around and slow down at its destination. Fusion or annihilation-drive ships will probably do this. They will apply power all the time, speeding up and then slowing down.

In this simple case, all the important performance parameters can be expressed on a single graph. This one is drawn for the case when 90% of the starting mass is propellant. Jet velocity and starting acceleration are the graph scales. Distances for several bodies are shown. Mars varies greatly; I used 150 million km. Trip times and specific power levels are also shown. "Specific power" expresses how much power the ship generates for each kilogram of its mass, i.e., its total power divided by its mass. The propellant the ship will carry is not included in the mass value.

An example: Suppose your ship can produce 100 kW/kg of jet power. You wish to fly to Jupiter. Where the 100kW/kg and Jupiter lines cross on the graph, read a jet velocity of 300,000 m/sec (Isp = 30,000) and an initial acceleration of nearly 0.01g. Your trip will take about two months.

The upper area of the graph shows that high performance is needed to reach the nearest stars. Even generation ships will need, in addition to very high jet velocities, power on the order of 100 kW/kg. The space shuttle orbiter produces about 100 kW/kg with its three engines. The high power needed for starflight precludes its attainment with means such as electric propulsion.

about 0.006% of the primordial mixture is helium-3, according to astrophysicist Fred Hoyle. This seemingly small amount is enough to make scooping up Jupiter's atmosphere from an orbiting ship very profitable. The power available, from fusion of scooped-up helium-3 and deuterium, to a "Jupiter-scoopiter" would be something like thirty times that needed to maintain its orbit in the face of the scooping drag.

The deuterium/helium-3 reaction produces relatively few neutrons, and would greatly reduce neutron heating. The deuterium/deuterium reaction produces about one-fourth as many neutrons as the deuterium/tritium reaction, but is roughly ten times harder to ignite. The boron/hydrogen reaction produces no neutrons, but may be too hard to ignite ever to be used in a magnetic bottle reactor. Deuterium, boron, and hydrogen, of course, are all plentiful.

Star-drive reactors will have to be far more powerful than any yet conceived. At very high powers, new reactor types might be contemplated. One such is a pellet-collision scheme. Electromagnetic accelerators would fire small pellets of boron hydride at one another, at speeds of hundreds of kilometers per second. The high temperatures and densities produced by the collision would ignite the fusion reaction. A magnetic rocket nozzle would direct the jet. Maybe the now-defunct program of the 1950s to develop boron hydride missile fuels was strangely prescient!

Still more powerful star-drives might be contrived from matter-antimatter engines. Anti-hydrogen can be made from positrons (anti-electrons) and anti-protons. The former were discovered in the 1930s, and anti-protons were first isolated in 1955. Plans are now being discussed for containment of up to a trillion anti-protons in a storage ring for physics experiments. One fine day, we will undoubtedly be able to create and store anti-hydrogen, or anti-helium, or whatever. This, of course, will be sporty, inasmuch as any form of antimatter will annihilate ordinary matter upon contact! We will have to find a storage scheme that somehow suspends the antimatter fuel without its coming into contact with anything.

The payoff is that the matter-antimatter reaction will be able to convert mass into energy at more than 50% efficiency, whereas fusion is less than 1% efficient. This is the best star-drive that presently known physics can prescribe: an annihilation engine, powered by a matter-antimatter reaction. Such a drive might approach lightspeed. Built large enough, an annihilation engine might attain high thrust and yield a star-drive able to shorten star trips by relativistic time dilation. We are now speculating rather than forecasting, of course. No one has published a design for an annihilation star-drive, but this is clearly the ultimate that presently understood physics has to offer.

What might a starship be like? Probably rather like one of Gerard K. O'Neill's space colonies. Imagine that we have found a planetary system orbiting a suitable nearby star. We know there are some Earth-size planetary bodies, but we don't know much about them. Colonies and bases on the moon, Mars, and some of the moons of Jupiter and Saturn have taught us to survive on inhospitable worlds. We are confident that if we can get there, we can survive.

Survival, however, depends on perpetuation of a high-technology society. We must take along enough skills and knowledge to operate, repair, rebuild, replicate, and improve all of the machinery and equipment we need to jive and prosper, perhaps on a moonlike body. And our raw materials are likely to be undifferentiated rocks, not high-grade ores. How many people do we need to embody all the skills? Remember, no help from home! Today, my guess is about ten thousand. In the future, developments in robotics might reduce the number, but it will still be large.

One thing we will surely do is listen before we leap. We would not wish to arrive in a distant planetary system only to find we were not welcome. If indigenous intelligent civilization is present, careful listening with large radio telescopes will reveal it by detecting "leakage" of electromagnetic signals such as radio or radar, even if these are not meant for interstellar communication. And we should not chance that an apparent welcome at departure time will last until arrival. Imagine a Japanese expedition to the United States departing in 1903 for arrival forty years later. Things change.

Our starship will probably gross on the order of a million metric tons and be something like a thousand meters (eleven football fields) in length. A triangular wedge shape similar to the empire ships in *Star Wars* will minimize the damage from hitting anything at high speed. Let's hope we don't hit anything bigger than a grain of sand. At 0.1c, a collision with a pea-sized pebble would release as much energy as a small atom bomb.

The star-drive will provide artificial gravity, perhaps a tenth of a "g," onboard the ship. Our program will be to accelerate until we are halfway there, then turn around and decelerate the rest of the way. If we are to travel ten light-years, accelerating at 0.1g, our maximum speed will be about 0.15c and the trip will take 130 years. If our star-drive has a specific impulse of a million seconds and a billion newtons (about 200 million pounds) of thrust for our 0.1g, its power will be several billion megawatts. This is about a thousand times the present total energy consumption of the United States. Such huge numbers sound like they come from a government budgeteer, but these are the awesome statistics of starship engineering as we now know it. Ah, for even the glimmer of an idea how to pull off the spacewarp trick!

If we have some ideas as to "how," then we are surely allowed to ask "when." The "when" of starflight may be more an economic question than a technical one. If we had the technology to build it, our million-ton ship would likely cost a trillion dollars. Such a proposal will never get by Senator Proxmire.

Further, when starships first become possible, the expected trip time will be twenty to one hundred years. (Figure 2.) The return on an investment, i.e., to Earth, will be nil. What will motivate the spending of huge sums on such an enterprise?

Our society today is willing to make huge investments only if a near-term payback is expected or if some urgent need, such as "war on poverty" or national security, is served. What the U.S. government spends on social causes, including social security, would pay for a trillion-dollar starship in three years; what it spends on defense, in about five. At the present NASA spending rate, however, it would take 160 years to pay for the starship!

This shortsighted attitude toward investment in scientific or cultural advancement seems to stem from a "Harvard Business School" outlook. If one evaluates investments with a discount rate of ten percent or more, anything that takes more than ten or fifteen years to mature is worthless. (Yet, for some reason, we keep on having children.) As usually practiced, the Harvard Business School approach is not even good economics, since it ignores indirect benefits such as technical advancements; these are often of more value than the direct benefits.

Ever since discounting became popular for investment analysis in the mid-1960s, our technological and cultural advancement has been slowed. But it was not always thus. The great cathedrals of Europe were constructed over generations. The U.S. interstate highway program has taken about thirty years. One wonders how long the pyramids of Egypt took, and what percent of Egypt's gross national product they consumed.

We shouldn't be too pessimistic. Starship technology is probably at least fifty years in the future. It may be as far in the future as the clipper ships are in the past. Our ability to fund epic projects continues to advance as our economy grows. At an annual growth rate of three percent, our economy will multiply by a factor of twenty in 100 years. A starship project will not be too staggering to contemplate. It will, in fact, be less of a fiscal challenge than was Project Apollo in 1961.

Some will argue about "limits to growth" and all that. Indeed, the finiteness of Earth's resources will force us into space if we are to maintain economic progress. The industrialization and colonization of the solar system will surely bring about great cultural changes in our society. It's difficult to guess exactly what they will be, but I would bet they will improve the social acceptability of a starship project.

Now that we can begin to sketch the outlines of engineering starships, we are naturally drawn to questions about alien interstellar travelers. In the 1950s, and especially in the 1960s, the popular scientific view held that the universe is teeming with life, presumably intelligent. Radio telescopes listened for artificial signals from planetary systems of other stars. Discovery was announced of one or more planets orbiting Barnard's Star. Primitive life on Mars was almost accepted as fact. Enrico Fermi, anticipating a later view by about twenty years, asked, "Where are they?"

The United States once regarded the Atlantic Ocean as a protective barrier. Similarly, the great gulfs between the stars have been viewed until recently as a barrier to starflight. Discovery of intelligent life elsewhere in the universe, even though of enormous philosophical and scientific importance, was seen strictly as an academic matter and not one of practical concern.

In the last few years, all this has begun to change. The key development has been Gerard O'Neill's proposals for self-sufficient space colonies. Fitted with a fusion propulsion system, such a colony might become a "generation starship," i.e., a ship capable of reaching the stars in a few generations of life aboard. Starflight by this means is no longer merely fanciful science fiction but technically conceivable.

Has intelligent life arisen somewhere else in the galaxy? Even if it spread only at a speed of 0.01c, such an intelligence could colonize the entire galaxy in two to three million years. This is about the length of time that manlike hominids have lived on Earth, an exceedingly short time in the history of the galaxy. A speed of 0.01c is one one-hundredth the speed of light. A fusion-propelled ship should easily exceed 0.01c; 0.04 to 0.05c is a more likely estimate. If we presume that hundreds of years are spent at each new stellar beachhead, to regroup and gather strength for the next leap across the dark reaches of interstellar space, a spreading velocity of 0.01c is quite believable. Indeed, where are they?

Recognition that construction of a starship may be possible with foreseeable extensions of our technology, coupled with a distinct lack of alien visitors, seems to be swinging scientific opinion toward doubting the occurrence of life elsewhere in the universe. Perhaps we are alone! Recently it has been speculated (*Science News*, June 20, 1981) that supernova explosions within the galaxy, or its central core, have prevented intelligent life from developing by spraying lethal doses of radiation throughout space. Earth is imagined to have escaped by fortuitously having been in the protective envelope of a galactic spiral arm during each lethal period.

Indeed, "Where are they?" If we can, why haven't "they"—some other intelligent civilization—achieved starflight? And why aren't they here or, at least, why haven't they been here and left some record? Of course, many people believe they *are* (UFOs) or have been ("ancient astronauts"). But the evidence for either of these views is hardly convincing. At least, I am not convinced!

One can think of dozens of reasons why an advanced civilization, spreading through the galaxy, would have bypassed Earth. Even so, current scientific opinion is veering away from the "teeming with life" views of the 1960s.

If starships are big and expensive, as I suspect, it is likely that they don't pass this way very often, perhaps every few million years. And why didn't "they" land here and colonize? "Star Trek" had an answer: the "prime directive." Don't interfere with indigenous life or cultures. Maybe the spaceship phase of technical and cultural development is only transitory. How about teleportation? Or some sort of super-sensory means of communication such that there simply is no reason to travel? For that matter, maybe "they" will land tomorrow. The fact that we do not have alien visitors proves neither (a) that aliens don't exist, nor (b) that star travel is impossible.

I don't suppose I can avoid some sort of speculation on the spacewarp trick. Of course, it has no basis in known science. And scientists are forever acting as if they know all there is to know about the universe. That was certainly true about 1890. There was actually serious scientific opinion back then that no more basic science remained to be discovered. We knew all there was to know! There have been, since then, about three revolutions in physics. Relativity and quantum mechanics are the biggies, and solid-state physics has had a profound impact on almost everyone's life. The pocket calculator lying on my desk could only have been explained as "magic" as recently as the 1940s.

We are apparently in the midst of another revolution in physics. The "gauge theories" of particle physics are coming close to Einstein's dream of a unified field theory, a single theory that explains all of the forces of nature. They even offer an explanation as to why antimatter is apparently very scarce in the universe.

These theories, in their present state, with all the stuff about colored and charmed quarks, are mainly mechanistic. They are like Kepler's elucidation of planetary motion. He discovered that planets move in ellipses, and that an imaginary line between the planet and the sun sweeps equal areas in equal times, and so forth. But it remained for Newton to propose gravity as an explanation and interpretation of Kepler's facts.

Similarly, the trappings of relativity, the constancy of lightspeed, the Lorentz-Fitzgerald contraction (it isn't named for Einstein!) were mostly known before Einstein offered his profound philosophical interpretation, with its awesome insight to the equivalence of matter and energy.

We await another great philosopher-interpreter to transform the multitude of discoveries in particle physics and cosmology of the last twenty or so years into a new synthesis, a new and deeper explanation of the mysteries of nature. Out of his (her?) work may emerge the secret of star travel without strain and pain. Otherwise, we will just have to use brute force and high-energy propulsion.

If our human civilization does survive its near-term crises of militarism, nationalism, and resource exhaustion, we will soon move out into the solar system. And we will eventually build starships. Our first starship will not be our last. ". . . several bright sources such as quasar 3C273 are famous for having high-speed jets protruding out one side."—*Science News*, August 15, 1981. It sort of makes you wonder.